AF587394

Waste and Waste Management

Development of End-Markets for Recycled Construction and Demolition Waste Resources in Australia
Salman Shooshtarian, PhD (Editor)
Tayyab Maqsood, PhD (Editor)
Savindi Caldera, PhD (Editor)
Tim Ryley, PhD (Editor)
Peter SP Wong, PhD (Editor)
2023. ISBN: 979-8-88697-956-5 (Hardcover)
2023. ISBN: 979-8-89113-121-7 (eBook)

Fuel Briquettes Made of Carbon-Containing Technogenic Raw Materials
Nina Buravchuk, PhD (Editor)
Olga Guryanova (Editor)
2023. ISBN: 979-8-88697-907-7 (Softcover)
2023. ISBN: 979-8-88697-944-2 (eBook)

Water Supply and Drainage for Buildings
Syed Azizul Haq (Editor)
2023. ISBN: 979-8-88697-607-6 (Hardcover)
2023. ISBN: 979-8-88697-761-5 (eBook)

Municipal Solid Waste Management and Improvement Strategies
Adam Fitz (Editor)
2023. ISBN: 979-8-88697-720-2 (Softcover)
2023. ISBN: 979-8-88697-756-1 (eBook)

What is Biodegradation and Why it Matters
Robert T. Howard (Editor)
2022. ISBN: 978-1-68507-933-8 (Hardcover)
2022. ISBN: 978-1-68507-959-8 (eBook)

More information about this series can be found at
https://novapublishers.com/product-category/series/waste-and-waste-management/

Waste and Waste Management

Ken Ykstra
Editor

Nuclear Waste

Safety, Security and Clean-Up

NOTICE TO THE READER

Library of Congress Cataloging-in-Publication Data

ISBN: 979-8-89113-719-6 (Hardcover)
ISBN: 979-8-89113-766-0 (eBook)

Published by Nova Science Publishers, Inc. † New York

Contents

Preface

Nuclear weapons production and energy research activities generated large amounts of radioactive contamination and hazardous waste at commercial enterprises. As a result, contamination of soil, groundwater, and structures occurred at sites across the country, posing potential risks to human health and the environment. This book provides information about the safety and disposal of the waste products and the efforts to clean up the contamination.

Chapter 1

Cleaning up Communities: Ensuring Safe Storage and Disposal of Spent Nuclear Fuel*

Committee on Energy and Commerce

Thursday, June 13, 2019
House of Representatives,
Subcommittee on Environment and Climate Change,
Committee on Energy and Commerce,
Washington, DC.

The subcommittee met, pursuant to call, at 10:03 a.m., in room 2322, Rayburn House Office Building, Hon. Paul Tonko (chairman of the subcommittee) presiding.

Present: Representatives Tonko, Clarke, Peters, Barraga´n, Blunt Rochester, Soto, DeGette, Matsui, McNerney, Ruiz, Dingell, Pallone (ex officio), Shimkus (subcommittee ranking member), McKinley, Johnson, Long, Flores, Carter, Duncan, and Walden (ex officio).

Staff Present: Adam Fischer, Policy Analyst; Waverly Gordon, Deputy Chief Counsel; Rick Kessler, Senior Advisor and Staff Director, Energy and Environment; Brendan Larkin, Policy Coordinator; Tuley Wright, Energy and Environment Policy Advisor; Mike Bloomquist, Minority Staff Director; Adam Buckalew, Minority Director of Coalitions and Deputy Chief Counsel, Health; Peter Kielty, Minority General Counsel; Mary Martin, Minority Chief

---

* This is an edited, reformatted and augmented version of Hearing before the Subcommittee on Environment and Climate Change, of the Committee on Energy and Commerce, House of Representatives, One Hundred Sixteenth Congress, First Session, Serial No. 116–45, dated June 13, 2019.

In: Nuclear Waste
Editor: Ken Ykstra
ISBN: 979-8-89113-719-6

Counsel, Energy and Environment and Climate Change; Brannon Rains, Minority Legislative Clerk; and Peter Spencer, Minority Senior Professional Staff Member, Environment and Climate Change.

Mr. TONKO. The Subcommittee on Environment and Climate Change will now come to order. I recognize myself for five minutes for the purpose of an opening statement.

Opening Statement of Hon. Paul Tonko, a Representative in Congress from the State of New York

Politics of nuclear waste disposal are unquestionably difficult. In 1982, Congress passed the Nuclear Waste Policy Act directing the Department of Energy to remove spent nuclear fuel from commercial nuclear power plants in exchange for certain fees and transported to a permanent geologic repository beginning no later than January 31 of 1998; 1998 has come and gone. And year after year, we continue to debate how Congress can help break the impasse in which we currently find ourselves.

Today, there are over 70,000 metric tons of waste, which is expected to grow significantly in the decades to come. We are also dealing with more and more reactors shutting down, many of which are decommissioning early.

I take our Nation's nuclear waste challenges seriously. Today, there is not an easy or clear solution. But while we fail to make progress, American taxpayers continue to make payments from Treasury's judgment fund. There are many members on this committee on both sides of the aisle that would like to see a fair outcome that acknowledges these challenges, finds workable solutions, and protects American taxpayers.

And I want to give credit to Mr. Shimkus for his tireless efforts on this issue. I appreciate his commitment to helping communities dealing with waste and seeking to protect taxpayers from future need to make payments from the Treasury.

Today, the subcommittee will consider three bills which take different steps to address our Nation's nuclear waste issues. First, H.R. 2699, the Nuclear Waste Policy Amendments Act of 2019, introduced by Mr. McNerney and Mr. Shimkus. It is very similar to H.R. 3053 from the 115th Congress, which passed this committee and the House with bipartisan support. The bill makes a number of updates to the Nuclear Waste Policy Act. H.R. 3136, the STORE Nuclear Fuel Act of 2019, introduced by Ms. Matsui, directs the

Secretary of Energy to establish an interim storage program. And then, finally, H.R. 2995, the Spent Fuel Prioritization Act of 2019, introduced by Congressman Mike Levin, which would require the Secretary of Energy to prioritize the removal of spent nuclear fuel from decommissioned nuclear sites in areas with large populations and high seismic hazard.

I doubt any piece of legislation alone will solve our waste challenges, but I do believe that we need to be considering all options for disposal in an effort to find the safest and, indeed, most cost-effective way to move forward.

Today's panel attempts to cover many different and critical perspectives, and I look forward to the discussion.

[The prepared statement of Mr. Tonko follows:]

Prepared Statement of Hon. Paul Tonko

The politics of nuclear waste disposal are unquestionably difficult.

In 1982, Congress passed the Nuclear Waste Policy Act directing the Department of Energy to remove spent nuclear fuel from commercial nuclear power plants in exchange for certain fees, and transport it to a permanent geologic repository beginning no later than January 31, 1998.

1998 has come and gone, and year after year we continue to debate how Congress can help break the impasse in which we currently find ourselves.

Today, there are over 70,000 metric tons of waste, which is expected to grow significantly in the decades to come. We are also dealing with more and more reactors shutting down, many of which are decommissioning early.

I take our nation's nuclear waste challenges seriously.

Today there is not an easy or clear solution.

But while we fail to make progress, American taxpayers continue to make payments from Treasury's Judgment Fund.

There are many Members on this Committee, on both sides of the aisle, that would like to see a fair outcome that acknowledges these challenges, finds workable solutions, and protects American taxpayers.

And I want to give credit to Mr. Shimkus for his tireless efforts on this issue. I appreciate his commitment to helping communities dealing with waste and seeking to protect taxpayers from future need to make payments from the Treasury.

Today, the Subcommittee will consider three bills which take different steps to address our nation's nuclear waste issues.

H.R. 2699, the Nuclear Waste Policy Amendments Act of 2019, introduced by Mr. McNerney and Mr. Shimkus, is very similar to H.R. 3053 from the 115th Congress, which passed this Committee and the House with bipartisan support. The bill makes a number of updates to the Nuclear Waste Policy Act.

H.R. 3136, the STORE Nuclear Fuel Act of 2019, introduced by Ms. Matsui, directs the Secretary of Energy to establish an interim storage program.

And finally H.R. 2995, the Spent Fuel Prioritization Act of 2019, introduced by Congressman Mike Levin, would require the Secretary of Energy to prioritize the removal of spent nuclear fuel from decommissioned nuclear sites in areas with large populations and high seismic hazard.

I doubt any piece of legislation alone will solve our waste challenges, but I do believe we need to be considering all options for disposal in an effort to find the safest and most cost-effective way to move forward.

Today's panel attempts to cover many different and critical perspectives, and I look forward to the discussion.

With that, I yield the remainder of my time to Mr. McNerney or Mrs. Dingell.

Mrs. DINGELL. Thank you, Mr. Chairman.

I am going to be very brief. But I just really want to thank you for having this hearing. We are going to hear all the reasons why it continues to be an issue.

But with more than 20 percent of the fresh water in the world being in the Great Lakes, the threat of nuclear waste being stored there continues to be an enormously frightening issue for those of us in the Midwest but should be for everyone in this country. And I will yield back.

Thank you.

Mr. TONKO. The gentlewoman yields back. You are most welcome.

And now the Chair will recognize Mr. Shimkus, ranking member for the Subcommittee on Environment and Climate Change, surrounded by several boxes, for five minutes for his opening statement.

Opening Statement of Hon. John Shimkus, a Representative in Congress from the State of Illinois

Mr. SHIMKUS. Thank you, Mr. Chairman.

I am glad so many of my colleagues were able to join us today, because every time we get a chance to address this, we get to continue the educational process.

So, before I address the three bills today, I do want to explain what these stacks of science beside me are and why they matter in this debate.

Together, these represent the Federal Government's scientific and technical case for the permanent geological repository we are required by law to build and operate. This is a product. The larger of the two is the Department of Energy's 16 volume applications submitted to the Nuclear Regulatory Commission in 2008 to license a permanent repository for nuclear waste to Yucca Mountain.

So, my colleagues, this is $10 billion of, actually, those of you who live in nuclear States, that you have paid for because of the rate-based funding stream.

This is a product of 20 years and more than $10 billion of scientific research by eight of our national labs, our National Academies of Science and Engineering, the U.S. Geological Survey as well as DOE staff and contractors. It demonstrates that DOE can safely build and operate the repository in compliance within NRC's regulations.

This smaller stack, I point my colleagues to this big but smaller stack, is the NRC's five-volume analysis of this application. This is a product of NRC's technical staff and experts in geochemistry, hydrology, climatology, structural geology, volcanology, seismology, health physics, as well as chemical, civil, mechanical, nuclear, mining materials, and geological engineering.

Their review of DOE's application found—and the NRC is an independent agency of the United States Government. Their review of DOE's application found that, as proposed, a repository at Yucca Mountain would safely contain spent nuclear fuel and high-level waste for, get this, one million years. I trust the work done by the world-class scientists and engineers who produced these reports. I believe their work can stand up to the scrutiny of the skeptics if put to the test. Those skeptics, some of whom we will hear from today, don't believe the science. They claim Yucca Mountain is unsuitable based upon their own studies. Yet they are unwilling to make their contentions before the Atomic Safety and Licensing Board Panel of judges who are themselves scientists and engineers. That is the appropriation debate we are having. The skeptics who do not believe in science are unwilling to take their science and have it debated in front of this board.

Members have at their desk a packet with the NRC backgrounder on the licensing process—it is in the brown folder at your desk—a State-specific fact sheet, a map of the 121 sites in 39 States, a 1-pager addressing transportation concerns, the Peters-Duncan letter to appropriators seeking funding for the licensing, a 1-pager on the cost of doing nothing, which is $2 million a day, a chart showing the funding, or lack thereof, of appropriations since 1997.

So, I would ask my colleagues to keep that in mind today as we consider the stalemate, we find ourselves in. Ask yourself why those who oppose Yucca Mountain on supposedly scientific grounds would object to having their day in court.

With that said, the three bills we are here to talk about all reflect sincere efforts to address concerns that are arising out of our present impasse. And as we examine these proposals, we must remember the broader framework of our Nation's nuclear waste policy. This framework, funded almost entirely by ratepayers, is based upon a system that ensures that there will be a permanent repository. That is the focus of the fees collected in the Nuclear Waste Fund and the focus of the contract the utilities signed with the Department of Energy to eventually dispose of the spent fuel.

It is the point of the taxpayer spending for disposing of defense waste. You cannot take shortcuts here, whether it is with allowing the scientific adjudication to go forward or develop a system that accelerates the transportation of stranded fuel from decommissioned sites.

Shortcuts lead to dead ends, and I am concerned that the proposed measure that focused solely on interim storage without integrating it into a permanent system for disposal may sound good and expedient, but they will not work, and they may actually harm ratepayers and taxpayers in the long run.

I have a few more paragraphs, Mr. Chairman, but in lieu of time, I will submit those for the record.

And I will yield back.

[The prepared statement of Mr. Shimkus follows:]

Prepared Statement of Hon. John Shimkus

Before I address the three bills before us today, I want to explain what these two stacks of science beside me are and why they matter in this debate.

Together, these represent the Federal Government's scientific and technical case for the permanent, geological repository we are required by law to build and operate.

The larger of the two is the Department of Energy's 16 volume application, submitted to the Nuclear Regulatory Commission in 2008, to license a permanent repository for nuclear waste at Yucca Mountain.

This is the product of 20 years and more than $10 billion of scientific research by eight of our National Labs, the National Academies of Science and Engineering, the U.S. Geological Survey, as well as DOE staff and contractors. It demonstrates that DOE can safely build and operate the repository in compliance with NRC's regulations.

The smaller stack is the NRC's five volume analysis of DOE's application. This is the product of NRC's technical staff and experts in geochemistry, hydrology, climatology, structural geology, volcanology, seismology and health physics, as well as chemical, civil, mechanical, nuclear, mining, materials and geological engineering. Their review of DOE's application found that, as proposed, a repository at Yucca Mountain would safely contain spent nuclear fuel and high-level waste for one million years.

I trust the work done by the world-class scientists and engineers who produced these reports. I believe their work can stand up to the scrutiny of the skeptics if put to the test.

Those skeptics, some of whom we'll hear from today, don't believe this science. They claim Yucca Mountain is "unsuitable" based upon their own studies, yet they're unwilling to make their contentions before an Atomic Safety and Licensing Board panel of judges who are themselves scientists and engineers.

Members have at their desk a packet with (1) an NRC backgrounder on the licensing process, (2) a state specific fact sheet, (3) a map of the 121 sites in 39 states, (4) a one pager addressing transportation concerns, (5) the Peters-Duncan letter to appropriators seeking funding for the licensing, (6) a one pager on the cost of doing nothing, and (7) a chart showing the funding (or lack thereof) appropriated since 1997.

So I would ask my colleagues to keep that in mind today as we consider the stalemate we find ourselves in. Ask yourself why those who oppose Yucca Mountain on supposedly scientific grounds would object to having their day in court.

With that said, the three bills we're here to talk about all reflect sincere efforts to address concerns that are arising out of our present impasse. And as we examine these proposals, we must remember the broader framework of our nation's nuclear waste policy.

This framework, funded almost entirely by ratepayers, is based upon on a system that ensures there will be a permanent repository. That is the focus of the fees collected in the Nuclear Waste Fund, and the focus of the contracts the utilities signed with the Department of Energy to eventually dispose of their spent fuel. It is the point of the taxpayer spending for disposing of defense waste.

You cannot take shortcuts here. Whether it's with allowing the scientific adjudication to go forward or developing a system that accelerates the transportation of stranded fuel from decommissioned sites.

Shortcuts lead to dead ends. I'm concerned that proposed measures that focus solely on interim storage without integrating it into a permanent system for disposal may sound good and expedient. But they will not work, and they may actually harm ratepayers and taxpayers in the long run.

For example, H.R. 3136, introduced by my friend Ms. Matsui, provides for interim storage, including private interim storage; it provides for consent-based siting for interim storage facilities; and it requires that waste fees on utilities do not start until there is a licensing decision on a repository.

Similar measures were part of H.R. 3053 that we moved with a strong bipartisan vote of 340–72 through the House last Congress. And they are reflected in H.R. 2699, the bill Mr. McNerney and I introduced last month that we are also considering today. But there are defects with H.R. 3136.

First, it will be very difficult to gain acceptance for interim storage if it is not directly linked to permanent disposal. And second, if interim storage is authorized to be funded out of the Nuclear Waste Fund, it could effectively force operating reactors, and ratepayers to pay more in the future. The more money diverted from that fund to interim storage, the more ratepayers - or taxpayers-will have to pay in the future for permanent disposal.

In contrast, H.R. 2699 preserves the fund for the repository system and links progress on interim to completing the adjudication of the permanent repository. Only by showing the public the full information from this process can we build the trust in the sites, the decisions we make, and the eventual success of our nuclear waste program.

Thank you Mr. Chairman, I look forward to our discussion this morning.

Mr. TONKO. So the gentleman yields back.

The Chair now recognizes Representative Pallone, chairman of the full committee, for five minutes for his opening statement. Mr. Pallone.

Opening Statement of Hon. Frank Pallone, Jr., a Representative in Congress from the State of New Jersey

Mr. PALLONE. Thank you, Chairman Tonko.

It has been over 30 years since Congress last made significant changes to this law. Unfortunately, in that time, very little has been accomplished to address our Nation's need to safely store and dispose of the spent nuclear fuel that is a byproduct of electricity generation at nuclear power plants across the country.

Today, there are 121 communities across the country that have nuclear waste nearby. These communities are rightfully expressing frustration as more and more nuclear plants close, but there is no concrete solution to storage or disposal of this spent nuclear fuel. Whether it is a general safety concern or the desire of the committee to redevelop the land currently housing the spent fuel, we must find a path forward to begin the process of moving nuclear waste out of these communities.

At this hearing, we will be discussing three bills that take different approaches to addressing the spent nuclear fuel stalemate in our country. Representative McNerney and Ranking Member Shimkus have introduced H.R. 2699, an updated version of the legislation reported by the committee and passed by the House in the last Congress. I want to thank both of them for their leadership on this issue. In the last Congress, then-subcommittee Chairman Shimkus worked with us to address our concerns and incorporate interim storage language authored by Representative Matsui. That led to a successful effort in the House, and I look forward to continuing to work with him and Mr. McNerney on this issue.

We will also discuss H.R. 3136, the STORE Nuclear Fuel Act, introduced by Ms. Matsui. The bill establishes an interim storage program at the DOE which will allow for consolidated temporary storage of spent nuclear fuel with priority given to waste currently stored at decommissioned nuclear power plants. Authorizing interim storage will allow DOE to consolidate waste at one or two sites instead of 121 sites in communities across the country. And consolidated storage will ensure spent nuclear fuel is managed more safely and securely, in my opinion, while allowing communities with decommissioned plants to begin working towards redeveloping those sites.

Interim storage is the best near-term solution to stop the nuclear waste stalemate, and I commend Representative Matsui for her efforts and leadership on this issue.

And, finally, the committee will review H.R. 2995, the Spent Fuel Prioritization Act, introduced by Representative Mike Levin of California. This bill prioritizes the removal of spent nuclear fuel from decommissioned nuclear plants in areas with large populations and high seismic hazard.

So, once again, I appreciate the efforts of the bill sponsors and thank them for their leadership on this important issue.

[The prepared statement of Mr. Pallone follows:]

Prepared Statement of Hon. Frank Pallone, Jr.

It has been over 30 years since Congress last made significant changes to this law. Unfortunately, in that time very little has been accomplished to address our nation's need to safely store and dispose of the spent nuclear fuel that is a byproduct of electricity generation at nuclear power plants across the country.

Today, there are 121 communities across the country that have nuclear waste nearby. These communities are rightfully expressing frustration as more and more nuclear plants close, but there is no concrete solution to storage or disposal of the spent nuclear fuel. Whether it is a general safety concern, or the desire of the community to redevelop the land currently housing the spent fuel, we must find a path forward to begin the process of moving nuclear waste out of these communities.

At this hearing, we will be discussing three bills that take different approaches to addressing the spent nuclear fuel stalemate in our country.

Representative McNerney and Ranking Member Shimkus have introduced H.R. 2699, an updated version of the legislation reported by the Committee and passed by the House in the last Congress. I would like to thank them both for their leadership on this issue. Last Congress, then Subcommittee Chairman Shimkus worked with us to address our concerns and incorporate interim storage language authored by Representative Matsui. That led to a successful effort in the House and I look forward to continuing to work with him and Mr. McNerney on this issue.

We will also discuss H.R. 3136, the STORE Nuclear Fuel Act, introduced by Ms. Matsui. The bill establishes an interim storage program at the Department of Energy (DOE), which will allow for consolidated, temporary storage of spent nuclear fuel, with priority given to waste currently stored at decommissioned nuclear power plants. Authorizing interim storage will allow DOE to consolidate waste at one or two sites instead of 121 sites in communities across the country. Consolidated storage will ensure spent nuclear fuel is managed more safely and securely, while allowing communities with decommissioned plants to begin working toward redeveloping those sites. Interim storage is the best near-term solution to stop the nuclear waste stalemate, and I commend Representative Matsui for her efforts and leadership on this issue.

Finally, the Committee will review H.R. 2995, the Spent Fuel Prioritization Act, introduced by Representative Mike Levin. This bill prioritizes the removal of spent nuclear fuel from decommissioned nuclear power plants in areas with large populations and high seismic hazard.

Once again, I appreciate the efforts of the bill sponsors and thank them for their leadership on this important issue.

Mr. PALLONE.And I would like to yield the balance of my time now to Ms. Matsui.

Ms. MATSUI. Thank you, Mr. Chairman.

Finding a solution to managing the disposal of spent nuclear fuel has been a top priority of mine for many years, particularly as my district utility, the Sacramento Municipal Utility District, is one of the many across the country forced to play host to this dangerous radioactive material long after they committed to do so.

I think we all agree that this stalemate is unsustainable. The best and most pragmatic path forward involves a consolidated interim storage program that will engage with affected States and local governments through a consent-based process. That is why I have introduced the STORE Nuclear Fuel Act, which puts forward a plan that has historically garnered broad support.

The Federal Government has reneged on its promise to our constituents. A consolidated interim storage approach will allow the over 120 communities across the country to redevelop nuclear reactor sites that for many have been decommissioned for years. I believe this is one of the greatest energy challenges of our time. And I am grateful to the committee today for bring up my bill for discussion. Thank you. And I yield back.

Mr. PALLONE. Unless someone else wants my time, Mr. Chairman, I yield back.

Mr. TONKO. The gentleman yields back.

The Chair now recognizes Representative Duncan from South Carolina for five minutes.

Opening Statement of Hon. Jeff Duncan, a Representative in Congress from the State of South Carolina

Mr. DUNCAN. Thank you, Mr. Chairman.

The Energy and Commerce Committee has an enduring and strong bipartisan record supporting nuclear energy. Nuclear is a critical component of our Nation's energy system. It also has been vital to our national security powering the nuclear Navy and providing for our common defense. Not only is nuclear power an affordable and reliable energy source, it is also emissions-free. Any serious efforts to reduce emissions from energy production to address the effects of climate change must include the continued use and expansion of nuclear power. And there is great potential, if we get the policies right, to benefit from nuclear energy far into the future.

Over the past few years, we have moved legislation to lay the groundwork for advanced nuclear and to ensure more efficient regulation of the existing reactor fleet. We have explored policies that will ensure nuclear infrastructure for tomorrow, ranging from advanced small modular reactors like those under development by Oregon-based NuScale currently in NRC licensing to advance fuel systems for the next generation of reactors.

Yet, as we look forward, we have the responsibility to ensure that we implement the existing policies that address the issue of long-term storage for spent nuclear fuel and the defense legacy waste that the Federal Government has a responsibility for cleaning up.

This is no small matter. Thirty-five years ago, Congress enacted the Nuclear Waste Policy Act into law. This law was the culmination of decades of experience by the Federal Government to develop a policy to permanently dispose of high-level radioactive waste and commercial spent nuclear fuel.

Some of the material was created during the Manhattan Project and through the Cold War at the Hanford site, a vital national security facility located on the Columbia River.

Today, this nuclear material sits on a vibrant waterway waiting to be processed and transported to Yucca Mountain repository in the Nevada desert. The Nuclear Waste Policy Act also established a fee tied to the generation of nuclear energy to finance the cost of a multigenerational disposal program. Along with 33 other States, Oregon ratepayers and South Carolina ratepayers, fulfilled their financial obligation under the law and paid the Department of Energy over $160 million to dispose of commercial spent nuclear fuel. That was Oregon, and $1.5 billion from South Carolina.

As we all know, the Federal Government has been prevented from completing the licensing process for a permanent repository. The cost to the American taxpayer to pay for the Federal Government's delay in opening Yucca Mountain repository have more than doubled to $35 billion since 2009. And that figure continues to escalate rapidly as time goes on. Meanwhile, the Federal Government has been paying out nearly a billion dollars a year from the judgment fund for its failure to dispose of the waste.

Against this backdrop, Mr. Chairman, I appreciate your moving forward on examining legislative reforms that can help to restart this process. The Energy and Commerce Committee should continue to lead the effort to ensure the Federal Government meets its moral and fiduciary responsibility to clean up its defense waste and ensure the permanent safe disposal of spent nuclear fuel which sits at 121 sites around the Nation.

We made important strides in the last Congress to reform the fundamental statute to help to accelerate this complicated process. My friend and Republican leader of this subcommittee, John Shimkus, led the work in the House to pass the Nuclear Waste Policy Amendments Act by a vote of 340 to 72. Unfortunately, that effort fell short in the Senate.

But we know from the last Congress and from the strong bipartisan support both on this committee and in the House of that legislation, how a thoughtful, deliberate, legislative process produces good policy. I am pleased to see the past work has informed the current work, particularly in H.R. 2699, led by Mr. McNerney, which follows closely to H.R. 3053 from the last Congress. This bill provides for accelerating interim storage of waste without undermining the important system for the permanent disposal established in the underlying law. This represents the best path forward for getting the Nation to a licensing decision which is necessary for public confidence in our nuclear waste program no matter the outcome of that decision.

Thank you, Mr. Chairman, for taking the lead on this legislation, and I yield back.

[The prepared statement of Mr. Duncan follows:]

Prepared Statement of Hon. Jeff Duncan

The Energy and Commerce Committee has an enduring and strong bi-partisan record supporting nuclear energy. Nuclear is a critical component of our nation's energy system. It also has been vital to our national security, powering the nuclear navy and providing for our common defense. And there is great potential—if we get the policies right—to benefit from nuclear energy far into the future.

Over the past few years we've promoted new nuclear technologies that will provide reliable, emissions free power for our homes and factories, our industrial processes, that will help address climate risks. We've moved legislation to streamline regulations to lay the groundwork for advanced nuclear and to ensure more efficient regulation of the existing reactor fleet We've explored policies that will ensure a nuclear infrastructure for tomorrow-ranging from advanced small modular reactors like those under development by Oregon-based NuScale, currently in NRC licensing, to advanced fuel systems for the next generation of reactors.

Yet as we look forward we have responsibility to ensure we implement the existing policies to address the back end of the nuclear fuel cycle, and the defense legacy waste that the Federal Government has a responsibility for cleaning up.

Here too, Mr. Chairman, I think the Energy and Commerce Committee should continue its leadership to identify what is necessary for the Congress to ensure the Federal government meets is moral and fiduciary responsibility to clean up its defense waste and ensure the permanent, safe disposal of spent nuclear fuel.

We made important strides in the last Congress to reform the fundamental statute to help to accelerate this complicated process. That fell short in the Senate, but I remain hopeful that the record of work by this Committee will continue to inform policy efforts.

This is no small matter. 35 years ago, Congress enacted the Nuclear Waste Policy Act into law. This law was the culmination of decades of experience by the federal government to develop a policy to dispose of high-level radioactive waste and commercial spent nuclear fuel permanently.

Some of the material was created during the Manhattan Project and through the Cold War at the Hanford site, a vital national security facility located on the Columbia River about 40 miles north of my district. Today, this nuclear material sits on a vibrant waterway waiting to be processed and transported to the Yucca Mountain repository in the Nevada desert.

The Nuclear Waste Policy Act also established a fee tied to the generation of nuclear energy to finance the costs of a multi-generational disposal program. Along with 33 other states, Oregon ratepayers fulfilled their financial obligations under the law and paid the Department of Energy over $160 million to dispose of commercial spent nuclear fuel.

I've note in the past how the Trojan nuclear power plant, located in northwest Oregon, stopped producing electricity in 1993, with the expectation that DOE would begin to remove the spent fuel in 1998, as was required by law. That never happened, as we know, and since the plant's decommissioning activities were completed in 2007, only spent nuclear fuel remains stranded at the site, hampering any redevelopment efforts surrounding the site.

This example is repeated across the nation, with states and communities waiting for DOE to fulfill its obligations and dispose of the spent fuel.

As we all know, the Federal government has been unable to complete the licensing process for a permanent repository. The costs to the American taxpayer to pay for the federal government's delay in opening the Yucca Mountain repository have more than doubled to $35 billion since 2009 and that figure continues to escalate rapidly as time goes on.

Against this backdrop, Mr. Chairman, I appreciate your moving forward on examining legislative reforms that can help to restart this process. We know from the last Congress, and from the strong bi-partisan support both on this Committee and in the House, how a thoughtful and deliberate legislative process produces good legislation.

I'm pleased to see this past work has informed the current work, particularly in HR 2699, led by Mr. McNerney, which follows closely the H.R. 3053 from the last Congress.

This bill provides for accelerating interim storage of waste without undermining the important system for permanent disposal established in the underlying law. This represents the best path forward for getting the nation to a licensing decision, which is necessary for public confidence in our nuclear waste program, no matter the outcome of that decision.

Thank you, Mr. Chairman for taking the lead on this important issue.

Mr. TONKO. The gentleman yields back.

The Chair would like to remind Members that, pursuant to committee rules, all Members' written opening statements shall be made part of the record.

Now we go to introduction of our witnesses. And we thank them all for joining today on what is a very important topic.

First, we have Ms. Maria Korsnick, President and CEO of Nuclear Energy Institute. Welcome. Mr. Robert Halstead, Executive Director, State of Nevada, Office of the Governor, Agency for Nuclear Projects. And, again, welcome. Mr. Austin Keyser, Director of Political and Legislative Affairs With the International Brotherhood of Electrical Workers. We welcome you. And Mr. Geoffrey Fettus, senior attorney with Nuclear Climate and Clean Energy

Programs at the Natural Resources Defense Council. Again, welcome. And, finally, welcome to Mr. Lake Barrett, former Acting Director, Office for Civilian Radioactive Waste Management with the United States Department of Energy.

Again, thank you all for your time today.

And before we begin, I would like to explain the lighting system. In front of you are a series of lights. The light will initially be green at the start of your opening statement. The light will turn yellow when you have 1-minute remaining. Please begin to wrap up your testimony at that point. The light will turn red when your time has indeed expired.

At this time, the Chair will recognize Ms. Korsnick for five minutes to provide her opening statement. Thank you.

Statements of Maria G. Korsnick, President and CEO, Nuclear Energy Institute; Robert J. Halstead, Executive Director, State of Nevada, Office of the Governor, Agency for Nuclear Projects; Austin Keyser, Director, Political and Legislative Affairs, International Brotherhood of Electrical Workers; Geoffrey H. Fettus, Senior Attorney, Nuclear, Climate, and Clean Energy Program, Natural Resources Defense Council; and Lake Barrett, Former Acting Director, Office of Civilian Radioactive Waste Management, U.S. Department of Energy

Statement of Maria G. Korsnick

Ms. KORSNICK. Thank you, Mr. Chairman.

I am Maria Korsnick, President and CEO of the Nuclear Energy Institute. I appreciate the opportunity to testify today on the three pieces of legislation which seek to address and end the long overdue stalemate of disposing of our Nation's used nuclear fuel.

Nuclear energy is the largest and most efficient source of carbon-free electricity in the United States. Currently, we have 97 commercial nuclear power plants in 29 States, and they provide nearly 20 percent of America's electricity and more than half of our Nation's emissions-free electricity. These reactors are carbon-free workhorses essential to addressing climate change in any realistic manner.

That said, the advanced reactors of tomorrow and the current U.S. operating fleet are continually subject to reputational damage because Congress, for two decades now, has played politics with the issue of used fuel. It is vitally important that the U.S. remain a global leader in the commercial nuclear arena. And yet we are the only major nuclear Nation without a used fuel management program.

The U.S. nuclear industry has upheld its end of the bargain. At sites in 35 States around the country, commercial used fuel is safely stored and managed awaiting pick up by the Federal Government, which was scheduled in 1998. Additionally, the Nuclear Waste Fund, which was set up to finance the development of a national repository, currently has over $41 billion in its coffers, which has been contributed by electricity consumers and nuclear generation companies. Each year, over $1.5 billion more in interest accumulates in this fund. And, finally, each day that we don't have a solution cost American taxpayer $2.2 million in damages.

In recent years, this has come to over $800 million annually, and it has the unfortunate distinction of being the single largest liability paid out of the judgment fund year after year. It is time to solve this, and I am excited to talk about how that can be achieved. We need a durable used fuel program. Politicizing this issue has stymied progress for far too long. We must allow the science, not politics, to guide us forward. But let me be clear: congressional action is necessary.

First, we need an answer on the Yucca Mountain license application. DOE submitted the application to the NRC more than a decade ago, and Congress directed the NRC to issue a decision in 2012. This deadline, like too many, was missed because DOE, without basis, shut down the Yucca Mountain project. For the sake of the communities holding stranded commercial fuel wishing to redevelop their sites and others holding high-level defense waste, we must move forward and allow Nevada's technical concerns with Yucca Mountain to be heard by NRC's independent administrative judges. This will allow a licensing decision to be determined based on its scientific merits rather than politics.

Second, as the licensing process of Yucca Mountain moves forward, interim storage can play an important role in helping to move spent fuel away from the reactor sites. Moving interim storage forward in parallel with the Yucca Mountain project helps to alleviate State and local concerns that interim storage will become a de facto disposal facility. This point was, in fact, highlighted just last week by a letter from New Mexico Governor Luja´n Grisham.

That said, I am pleased interim storage is addressed in several of the bills that are being discussed today, and I strongly believe interim storage can be successful if moved in parallel with Yucca Mountain.

Finally, the nuclear industry and electricity consumers around the country have paid their fair share to address the back end of the fuel cycle. In fact, the Nuclear Waste Fund's annual investment income alone was $1.5 billion last year. Both H.R. 2699 and H.R. 3136 correctly understand the importance of not prematurely reimposing the nuclear waste fee, especially given the substantial balance and large investment interest which accrues annually.

The industry believes the fee should not be reinstated until, one, the annual expense for the program's ongoing projects exceed the annual investment income on the fund. And, two, the projected lifecycle cost demonstrates that the fee must be reinstated to achieve full cost recovery over the life of the program.

The fact that we are here today considering three pieces of legislation that address solving the used fuel stalemate is a positive step in the right direction. The industry sincerely appreciates the committee's awareness and motivation to find a durable solution. We look forward to continuing to work with each and every one of you to reach a bipartisan approach for the long-term management of the Nation's used fuel.

Thank you, and I look forward answering your question. [The prepared statement of Ms. Korsnick follows:]

Summary of Testimony, Nuclear Energy Institute, Maria Korsnick, President and Chief Executive Officer, Subcommittee on Environment and Climate Change, House Energy and Commerce Committee, June 13, 2019

Used nuclear fuel is stored safely and securely at sites in 35 states, and promising private initiatives are underway to develop consolidated storage facilities. But these measures are intended to be temporary until the federal government meets its obligation to develop a permanent solution.

Action by the federal government is long overdue. Utilities and their electricity customers have done their part, as their contributions have resulted in the $41 billion balance in the Nuclear Waste Fund. In addition, taxpayers have been saddled with the consequences of the federal government's inaction as more than $7 billion in damages has already been paid from the Judgment Fund and billions more in liability will continue to mount the longer action is delayed. And let us not forget about the communities near the facilities now used to store used fuel. Congress owes it to these communities to ensure science—not political whims—determines the fate of the Yucca Mountain repository, the nation's only authorized disposal option.

NEI urges Congress to take the following critical steps, which will put the U.S. on the path toward a viable used fuel management solution:

1. *Reach a Decision on Yucca Mountain:* The NRC has yet to decide whether it will grant the DOE's license application for the Yucca Mountain project. We support completing the Yucca Mountain license application proceeding. But to move forward, Congress must grant the NRC's and DOE's requests for funding to complete their duties.
2. *Authorize Consolidated Interim Storage:* Consolidated interim storage would enable DOE to move dry casks from nuclear plant sites to a centralized location where it can be more efficiently managed until a permanent repository is built. We support the development of a consolidated interim storage program in willing host communities and states in parallel with completing the Yucca Mountain licensing proceeding. Moving both programs forward in parallel will help to alleviate concerns that interim storage will become *de facto* disposal. This concern was highlighted just last week in a letter from New Mexico Governor Lujan Grisham.
3. *Reform the Nuclear Waste Fund and Fee Process:* Any legislation that becomes law must ensure a more equitable fee collections process and provide access to the Nuclear Waste Fund for its intended purposes. Congress should not allow DOE to impose unnecessary new fees on electricity consumers. It would be unfair to restart such fees until there is a showing, at a minimum, that (1) the annual expenses for the program's ongoing projects exceed the Fund's annual investment income and (2) the projected life-cycle cost demonstrates that additional fees are necessary to achieve full cost recovery over the life of the program.

We look forward to continuing to work with lawmakers to reach bipartisan consensus on the best approach for the long-term management of the nation's used fuel.

Testimony for the Record, Nuclear Energy Institute, Maria Korsnick, President and Chief Executive Officer, Subcommittee on Environment and Climate Change, House Energy and Commerce Committee, June 13, 2019

I am Maria Korsnick, President and Chief Executive Officer of the Nuclear Energy Institute (NEI).[1] I appreciate the opportunity to provide testimony on the Nuclear Waste Policy Amendments Act of 2019 (H.R. 2699), the Storage and Transportation of Residual and Excess (STORE) Nuclear Fuel Act of 2019 (H.R. 3136), and the Spent Fuel Prioritization Act of 2019 (H.R. 2995). Continuing the conversation on these issues is an important step to revitalize the federal used nuclear fuel program. It is important that Congress provide the U.S. Department of Energy (DOE) with clarifying authority to support private consolidated interim storage facilities and direction to move the Yucca Mountain application forward.

Used nuclear fuel is stored safely and securely at sites in 35 states, and promising private initiatives are underway to develop consolidated storage facilities. But these measures are intended to be temporary until the federal government meets its legal obligation to develop a permanent solution. Action by the federal government is long overdue. The failure of the federal government to implement the statutorily required used fuel management program has given the industry a black eye for far too long despite the fact that nuclear generation provides more than half of the nation's carbon-free electricity. Further, there are many advanced reactor designs being developed that can usefully be deployed in the U.S. in the near future to meet our clean energy needs. Burdening these promising technologies with the weight of a floundering used fuel management program unnecessarily and unreasonably

---

[1] NEI is responsible for establishing policy on issues affecting the commercial nuclear energy industry. NEI has about 300 members, including companies licensed to operate U.S. commercial nuclear power plants, nuclear plant designers, major architect/engineering firms, fuel cycle facilities, materials licensees, labor organizations, universities, and other organizations involved in the nuclear energy sector.

limits the tools we have to combat climate change at a time when we need every carbon-free generation option available.

Utilities and their electricity customers have done their part, as their contributions have resulted in the $41 billion balance in the Nuclear Waste Fund. In addition, taxpayers have been saddled with the consequences of the federal government's inaction as more than $7 billion in damages has already been paid from the Judgment Fund and billions more in liability will continue to mount the longer delay on the program continues. And let us not forget about the communities near the facilities in 35 states now used to store the used fuel. The citizens of those communities, and particularly those where there is no longer an operating plant, are an often an overlooked constituency with a significant stake in a fully functioning used fuel management program. Both they and the site owners are currently prevented from redeveloping the land on which these storage facilities sit. Congress owes it to the citizens in these communities to take whatever steps are necessary to get used fuel moving offsite.

The Importance of Nuclear Power to the United States

Nuclear energy is the largest and most efficient source of carbon-free electricity in the United States. Currently, 97 commercial nuclear power plants in 29 states provide nearly 20 percent of America's electricity and more than half of the emissions-free electricity. Because electricity generation from nuclear energy does not release carbon dioxide and other harmful air pollutants, by maintaining the domestic nuclear fleet, the United States will not have to choose between the health of its electric grid and the health of its citizens. Nuclear plants run 24 hours a day, 7 days a week producing power with unmatched reliability, and have the added benefit of 18-24 months of fuel on site. Nuclear plants are hardened facilities that are protected from physical and cyber threats, helping to ensure we have a resilient electricity system in the face of potential disruptions.

Nuclear energy facilities are essential to the country's economy and the local communities in which they operate. The typical operating plant generates $470 million each year in the sale of goods and services in the local community, and employs 700 to 1000 workers. Construction of a new nuclear plant provides in the range of 3500 jobs at peak periods. Collectively, the nuclear industry contributes about $60 billion every year to the U.S. economy, through supporting over 475,000 jobs and producing over $12 billion annually in federal and state tax revenues.

The Used Nuclear Fuel Stalemate

Unlike fossil fuel-fired power plants, which emit carbon dioxide and other air pollutants to the atmosphere, nuclear generation's primary byproduct is contained in the solid fuel it uses to produce electricity. After generating electricity for about five years, used nuclear fuel assemblies are removed from the reactor and safely stored initially in a concrete and steel fuel pool. When cool enough that the used fuel no longer needs to be stored underwater—typically several years after removal from the reactor—it can be transferred and stored in dry casks, which are large steel-reinforced concrete containers. Over the past three decades, industry has safely loaded and placed into storage over 3000 of these containers. All the used fuel produced by the U.S. nuclear energy industry in more than 50 years of operation would, if stacked end to end, cover a football field to a height of approximately 10 yards.

Used nuclear fuel is stored safely and securely at reactor and storage sites around the country, but onsite storage was never intended to be permanent. The Nuclear Waste Policy Act of 1982 (NWPA) codified DOE's obligation to dispose of used fuel generated by U.S. commercial nuclear power plants and the reciprocal obligation of plants owners and operators to offset disposal costs by paying fees into the Nuclear Waste Fund. To cement these obligations, the NWPA required plant owners/operators to enter into the legally binding Standard Contract with DOE. Pursuant to that contractual obligation, the owners and operators of nuclear plants— and the consumers of their electricity—have paid billions into the Nuclear Waste Fund. However, despite these massive investments, the federal government has fallen far short of meeting its end of the bargain as no tangible progress has been made towards developing a durable used fuel program.

In enacting the NWPA, Congress recognized that it was important to drive the government's action to complete the project by providing statutory deadlines by which significant milestones were to be met. Most prominently, Congress directed DOE to begin accepting used fuel by January 31, 1998. To help ensure this date was met, Congress amended the NWPA in 1987 to designate Yucca Mountain as the sole candidate for a repository. Despite this statutory deadline, by the mid-1990s, DOE made clear that it could not meet the 1998 deadline. Nonetheless, as statutorily required, DOE extensively evaluated the Yucca Mountain site before formally recommending moving forward with the repository in 2002. Congress again endorsed moving forward with Yucca Mountain and established a 90-day deadline for DOE to submit a construction authorization application to the U.S. Nuclear Regulatory Commission (NRC). Missing another deadline, DOE did not submit its application to the NRC until 2008, which triggered a 2012 deadline for the NRC to complete its review of the application. This was yet an additional deadline that was missed because in 2010, DOE—without basis—shut down the Yucca Mountain project in the middle of the NRC's application review and hearing process.

The failure to meet these deadlines—and the resulting harm to the industry, consumers, taxpayers, and local communities—has spurred literally dozens of lawsuits.[2] These lawsuits were necessary to protect the rights of generating companies and electricity consumers, and required the expenditure of countless resources that would have been better used elsewhere. Two of the more recent lawsuits are particularly relevant to the current used fuel stalemate.

First, the U.S. Court of Appeals for the D.C. Circuit in 2013 ordered DOE to reduce the Nuclear Waste Fund fee to zero until either the Yucca Mountain project is revived as required by the Nuclear Waste Policy Act or Congress enacts an alternative plan.[3] This decision squarely places the ball before Congress to fund the Yucca Mountain project or develop a comprehensive alternative disposal plan. In short, the Court made clear that DOE would not be permitted to start assessing fees from electricity customers unless tangible progress in made.

Second, also in 2013, the D.C. Circuit ordered the NRC to complete safety and environmental reviews of the Yucca Mountain license application.[4] While these reviews by the NRC's technical staff have since concluded that Yucca Mountain complies with all regulations, a final decision awaits an extensive formal hearing in which Nevada and others opposing the project can present

---

[2] *See, e.g.*, *Texas v. United States*, 891 F.3d 553 (5th Cir. 2018); *Nat'l Ass'n of Regulatory Util. Comm'rs v. DOE*, 680 F.3d 819 (D.C. Cir. 2012); *Dairyland Power Coop. v. United States*, 645 F.3d 1363 (Fed. Cir. 2011); *Pacific Gas & Elec. Co. v. United States*, 536 F.3d 1282 (Fed. Cir. 2008); *Yankee Atomic Elec. Co. v. United States*, 536 F.3d 1268 (Fed. Cir. 2008); *Alabama Power Co. v. DOE*, 307 F.3d 1300 (11th Cir. 2002); *Roedler v. DOE*, 255 F.3d 1347 (Fed. Cir. 2001); *Northern States Power Co. v. United States*, 224 F.3d 1361 (Fed. Cir. 2000); *Maine Yankee Atomic Power Co. v. United States*, 225 F.3d 1336 (Fed. Cir. 2000); *Northern States Power Co. v. DOE*, 128 F.3d 754 (D.C. Cir. 1997); *Indiana Michigan Power Co. v. DOE*, 88 F.3d 1272 (D.C. Cir. 1996).

[3] *National Ass'n of Regulatory Utility Comm'rs.v DOE*, 736 F.3d 517 (D.C. Cir. 2013).

[4] *In re Aiken County*, 725 F.3d 255 (D.C. Cir. 2013).

evidence and arguments challenging DOE's license application to NRC's independent administrative judges. But these hearings cannot proceed absent further congressional appropriations.

Actions to Address Used Fuel are Well Understood and Technologically Achievable

Used fuel is and can continue to be stored safely onsite or at consolidated interim storage facilities. Ultimately, however, a permanent disposal solution is needed. The consensus within the scientific and technical community engaged in used fuel management is that safe geologic disposal is achievable with currently available technology.[5] Yet the U.S. is the only major nuclear nation without a used fuel management program. To help the U.S. maintain its role as a leader in the nuclear arena, NEI urges Congress to implement the following critical steps, which will put the U.S. on the path toward a viable used fuel management solution:

Reach a Decision on Yucca Mountain

The NRC has yet to decide whether it will grant the DOE's license application for the Yucca Mountain project. We support finishing the Yucca Mountain license application proceeding. But to move forward with either Yucca Mountain or another site, Congress must grant the NRC's and DOE's requests for funding to complete their duties. H.R. 2699 would help move the Yucca Mountain licensing proceeding forward by establishing a 30-month deadline (which may be extended by 12 months if necessary) for the NRC to issue a final decision on DOE's construction authorization application. The NRC missed the original deadline because of funding shortfalls and the absence of a cooperative applicant. A new deadline would add certainty and reinforce Congress's mandate to make meaningful progress. With adequate funding and commitment by DOE, the deadline is achievable and gives the NRC a reasonable timeframe in which to consider and resolve the challenges to the application raised by Nevada and other stakeholders.

---

[5] Blue Ribbon Commission on America's Nuclear Future, Report to the Secretary of Energy § 4.3 (Jan. 2012).

Authorize Consolidated Interim Storage

Used fuel is being safely stored in robust pools and airtight containers called dry casks. A consolidated interim storage program would enable the DOE to move dry casks from nuclear plant sites to a consolidated interim storage facility where it can be more efficiently managed until a permanent repository is built. We support the development of a consolidated interim storage program in willing host communities and states in parallel with completing the Yucca Mountain licensing proceeding. As the Committee recognized last year when it overwhelmingly approved H.R. 3053, moving the consolidated interim storage program forward in parallel with the Yucca Mountain project helps to alleviate state and local concerns that interim storage will become a *de facto* disposal facility and will distract from repository development. New Mexico Governor Lujan Grisham's June 7 letter to DOE Secretary Perry and NRC Chairman Svinicki identified this very concern.[6]

Reform the Nuclear Waste Fund and Fee Process

As noted herein, because of contributions of the owners and operators of nuclear plants—and the consumers of their electricity—the Nuclear Waste Fund has a balance of more than $41 billion and each year over $1.5 billion in interest is added to this principle balance. Any legislation that becomes law must ensure a more equitable fee collections process and provide access to the Nuclear Waste Fund. Historically, congressional budgeting practices have prevented the use of this fund for its intended purpose. Meanwhile, taxpayers have been have shouldered with the more than $7 billion in damages for the federal government's inaction—an amount that grows by $2.2 million for every day the government does not act. The cost of not funding a solution is rapidly becoming greater than the cost of funding the program. NEI supports—and electricity customers and taxpayers deserve—granting DOE access to the Nuclear Waste Fund for its intended purpose without reliance on the annual appropriations process.

Perhaps more important, however, is that DOE not impose on electricity consumers unnecessary new fees. It would be unfair to restart such fees until there is a showing, at a minimum, that (1) the annual expenses for the

---

[6] Letter from New Mexico Governor Michelle Lujan Grisham, to DOE Secretary Rick Perry and NRC Chairman Kristine Svinicki at 1-2 (June 7, 2019) (claiming that "the absence of a permanent high-level radioactive waste repository creates even higher levels of risk and uncertainty around any proposed interim storage site" and that "given that there is currently no permanent repository for high-level waste in the United States, any interim storage facility will be an indefinite storage facility").

program's ongoing projects exceed the annual investment income on the NWF and (2) the projected life-cycle cost demonstrates that additional fees are necessary to achieve full cost recovery over the life of the program. We appreciate that the sponsors of both H.R. 2699 and H.R. 3136 recognize the need to prohibit premature imposition of the fee. Also in a good faith effort to protect electricity consumers, H.R. 2699 appropriately limits fee collection to an amount not greater than 90 percent of what is appropriate from the Nuclear Waste Fund. While short of full access to the Nuclear Waste Fund, this approach is a good first step.

Used Fuel Is a Political Problem, Not a Technical One

As my testimony hopefully demonstrates, the government inaction impeding completion of a durable and permanent solution for used nuclear fuel is caused by political, not technical obstacles. But with strong leadership in Congress and in the Administration, they can be overcome.

In charting a path forward, Congress should not allow the political will of one state to stymie progress on an important project that would benefit 35 other states. This is not to say that Nevada should have no say in this process or that Yucca Mountain should be constructed without a full and fair airing of the concerns raised by those opposing the project.

Nevada and other stakeholders with technical concerns should be given every opportunity to demonstrate their perspectives on whether Yucca Mountain should be granted a license to receive fuel and operate as our nation's repository. There are approximately 300 contentions admitted in the NRC licensing hearing on Yucca Mountain. Should funding be restored and those proceedings restarted, Nevada and others that oppose the project can make their case to NRC's independent administrative judges—allowing a licensing decision on Yucca Mountain to be determined based on its scientific and technical merits. Congress owes it to the communities around the country where used fuel is currently being stored to ensure that science—not political whims—determines the fate of the Yucca Mountain repository, the nation's only authorized disposal option.

Adding to the fair treatment of Nevada and local communities, H.R. 2699 would make it possible for the State to engage in discussions regarding federal benefits without waiving objections to the project. Thus, should the NRC deem Yucca Mountain safe and should the project move forward, the bill would ensure that, notwithstanding its opposition to the project, Nevada would be eligible to enter into a benefits agreement.

Conclusion

On behalf of NEI and its members, I wish to thank the bill's sponsors for reintroducing the Nuclear Waste Policy Amendments Act of 2019. Just last year the House passed a very similar version of that bill with an overwhelming 340-72 bipartisan vote that represented a first step of progress to solving this longstanding issue. Today, the Committee is considering three pieces of legislation to address this problem. The industry sincerely appreciates the Committee's deliberate effort to find a durable solution. We look forward to continuing to work with lawmakers to reach bipartisan consensus on the best approach for the long-term management of the nation's used fuel. We urge lawmakers to ensure that resulting legislation protects both electricity consumers and taxpayers.

* * *

Mr. TONKO. Thank you, Ms. Korsnick.

And next, Mr. Halstead, you are recognized for five minutes, please.

Statement of Robert J. Halstead

Mr. HALSTEAD. Thank you, Chairman Tonko, Ranking Member Shimkus, and members of the subcommittee. I appreciate the opportunity to be here this morning. I am Robert Halstead. I am executive director of the Agency for Nuclear Projects, which is part of the Office of Governor Steve Sisolak. Governor Sisolak has made three points in his letter to the full Energy and Commerce Committee and to the chair and ranking member.

First, the State of Nevada opposes the Yucca Mountain project based on scientific, technical, and legal merits. Second, under the current law, only the Governor is empowered to consult with the Federal Government on matters related to siting a nuclear waste repository. And, third, Governor Sisolak has been fully briefed on the bills and specifically opposes H.R. 2699 and the Barrasso version in the Senate because it continues a central failing of the current Federal law, which was the selection of Yucca Mountain based on political science rather than Earth science. Governor Sisolak's letter is attached to my testimony. Attachment 2 explains in detail our specific concerns about H.R. 2699.

I want to thank the committee staff for helping me overcome the difficulties I had in filing my testimony for today. The rest of my comments summarize a few key points.

Yucca Mountain contradicts the foundational principle of geologic disposal, but the site itself, its geology and hydrology, not engineered barriers designed by humans must prevent the radioactive contamination of groundwater and the accessible environment for tens of thousands to one million years. Without engineered barriers, Yucca Mountain would inevitably contaminate an aquifer from which water is used for a variety of purposes, including drinking water, agriculture, food processing, and Native American religious ceremonies.

After three decades of DOE failures at Yucca Mountain, H.R. 2699 bets the farm on Yucca Mountain and doubles down on DOE. Both are bad bets. Yucca Mountain would have to survive Nevada's 218 admitted contentions and 30 new contentions if licensing restarts. It would likely be 20 years or more, if ever, before any spent fuel could be received at Yucca Mountain.

And as DOE's own studies show, walking away from Yucca Mountain and starting over with a repository in salt or shale would save tens of billions of dollars and likely be available sooner. DOE bungled the first repository program. They bungled the second repository program. They bungled the Oak Ridge monitored retrieval storage program in its pronuclear estate as Tennessee when now-Senator Lamar Alexander was Governor. He made the decision to cast the first veto under the Waste Policy Act. The nuclear program must be taken out of DOE if you want it to succeed.

Two other measures before your committee, the Levin and Matsui bills certainly are a good first start in addressing the problem of stranded nuclear fuel. That is a furtherance of the recommendations of the Blue Ribbon Commission on America's Nuclear Future. And, also, many people probably are not aware here yet. In 2018, the Western Interstate Energy Board has also adopted a resolution that would prioritize removal of stranded fuel. However, the funding provisions of H.R. 3136, like the funding provisions of 2699, need to be closely examined.

H.R. 2699 is not a good solution for stranded fuel because the licensing conditions would severely limit the amount of stranded fuel that could be accepted at the MRS. Yucca Mountain is not a good solution because the spent fuel at the shutdown reactors as all the reactors go into dry storage are being welded into storage canisters that are not compatible with DOE's license application. And the safety evaluation report that has been referred to by Representative Shimkus clearly says as a condition established by the NRC staff, that fuel can't be received without being repackaged.

Turning to consent-based siting, Nevada supports the Nuclear Waste Informed Consent Act, H.R. 1544, introduced by Representatives Titus, Horsford, and Lee of Nevada, and the companion bill, S. 649, to require a consent agreement between any State—but here we are particularly looking at Nevada—between DOE, the repository host State, affected counties, affected Indian Tribes prior to the construction of a repository. This would extend consent to the State of Nevada for Yucca Mountain. This approach is also going to need to be considered for the interim storage facilities in Texas and New Mexico.

We think the better vehicle for fixing the program is S. 1234, the Nuclear Waste Administration Act of 2019. Governor Sisolak concluded his letter with a pledge. If your committee is truly interested in fixing the Nation's broken nuclear waste program, my staff and I and Nevada's congressional delegation would be happy to meet with you and explore constructive alternatives. I hope the subcommittee will consider my testimony today as a first step in fulfilling Nevada's part of the Governor's pledge.

[The prepared statement of Mr. Halstead follows:]

Testimony of Robert J. Halstead, on Behalf of the State of Nevada, before the Subcommittee on Environment and Climate Change Committee on Energy and Commerce, United States House of Representatives, Hearing on "Cleaning up Communities: Ensuring Safe Storage and Disposal of Spent Nuclear Fuel," June 13, 2019

Chairman Tonko, Ranking Member Shimkus, and Members of the Subcommittee, thank you for the opportunity to participate in this hearing on the storage and disposal of spent nuclear fuel. I am Robert J. Halstead, Executive Director of the Nevada Agency for Nuclear Projects. The Agency is part of the Office of Governor Steve Sisolak. The Agency is vested by state law to carry out the duties and responsibilities imposed on the State of Nevada by the Nuclear Waste Policy Act (NWPA), as amended. The Agency's primary responsibility is to oversee and evaluate the U. S. Department of Energy's (DOE) programs to characterize, license, construct and operate a geologic repository at the proposed Yucca Mountain site in southern Nevada. I hire and supervise consultants and scientists who oversee DOE's activities involving the Yucca Mountain site. I have worked in the nuclear waste policy field for 40 years.

Governor Sisolak has stated his position on Yucca Mountain in a letter to the Chairman and Ranking Member of the Energy and Commerce Committee: "The State of Nevada opposes the project based on scientific, technical, and legal merits. I am totally opposed to any legislative effort to restart the Yucca Mountain project. As you and your members know, under the Nuclear Waste Policy Act of 1982, only the governor is empowered to consult with the federal government on matters related to the siting of a nuclear waste repository." Governor Sisolak's letter is Attachment 1 to my testimony.

Agency staff and consultants have thoroughly reviewed H.R. 2699, the Nuclear Waste Policy Amendments Act of 2019. Governor Sisolak concludes, based on our analysis, that "H.R. 2699 would do nothing to repair the central failing of the current federal law. In 1987, Congress substituted political science for earth science and selected Yucca Mountain in Nevada as the only site for repository development. H.R. 2699 would not only continue this failed policy; it would seriously weaken Nevada's current due process rights to challenge documented safety concerns and adverse environmental impacts in the legally-mandated licensing proceeding." Our revised comments are Attachment 2.

Yucca Mountain Is an Unsuitable Site for a Geologic Repository

The primary objective of HR. 2699 is to restart the forced siting of a repository at Yucca Mountain by requiring DOE and the U.S. Nuclear Regulatory Commission (NRC) to resume the adjudicatory portion of the NRC licensing proceeding under expedited rules and schedules. The State of Nevada opposes H.R. 2699 because it ignores the facts about Yucca Mountain. The site is unsuitable for a geologic repository because of its geology and hydrology, its proximity to military aircraft training and testing facilities, and its distance from existing mainline railroads. DOE's license application submitted to the NRC in 2008 cannot overcome the deficiencies of the site.

Nevada's opposition to DOE's license application is driven by technical deficiencies in DOE's repository engineering design. The proposed repository emplacement drifts would be located in fractured rock above the water table and would inevitably leak dangerous radionuclides into the groundwater, where they would be transported to an aquifer. Water from this aquifer is used for a variety of purposes, including drinking water, agriculture, food processing, and Native American religious ceremonies. DOE's proposed waste packages (the so-called transportation, storage and disposal containers or TADs), a critical element of DOE's license application, are now obsolete by utility standards. The spent fuel canisters from all U.S. reactors are a different design, and U.S. reactors are not going to adopt DOE's TAD design. DOE's proposed thermal loading scheme (intended to drive water away from the waste packages by heating the waste emplacement tunnels to the boiling point of water for a thousand years) has never been demonstrated and cannot be proved in licensing. DOE's proposed installation of tens of thousands of titanium drip shields (weighing 5 tons each) to protect the waste packages from corrosive infiltrating water relies on yet to be developed technologies, and may not prevent contamination even if perfectly installed. These highly speculative titanium drip shields are estimated to cost $8 billion to $20 billion.

DOE will bear a heavy burden of proof in the largest and longest licensing proceeding in the history of the NRC. Moreover, DOE will face extraordinary difficulties obtaining water permits from the State of Nevada for repository construction; obtaining water permits and rights-of-way for construction of a 300 mile railroad across a national monument and active grazing lands; and imposing over flight restrictions on military aircraft using the airspace above the repository surface facilities and adjacent lands currently used by the US Air Force.

Considering these technical and legal complexities, and the eight year lull in licensing activity, it would take 20-25 years before spent nuclear fuel or high-level radioactive waste could be received at the proposed Yucca Mountain repository. DOE and NRC preparation for resumption of licensing could require 18 months or more; the licensing proceeding for a construction authorization, and expected litigation, could require 96 months or more; construction of facilities and a 300-mile long railroad, licensing for receipt of spent nuclear fuel, and expected litigation, could require 120 months or more. Additional litigation is certain over NRC licensing rules; DOE program implementation; the Environmental Protection Agency (EPA) radiation protection standard; state water permits; railroad alignment selection, right-of-way acquisition and construction; and possibly other matters. These lawsuits could easily add another four to six years.

Nevada's estimates of time required are based on detailed analysis of previous NRC licensing proceedings and related lawsuits. Over these 20-25 years, the restarted repository program would likely require a minimum of $2 billion in average annual appropriations. Insufficient funding by Congress at any stage of licensing, construction and operational testing would delay the beginning of spent fuel receipt from reactors.

H.R. 2699 fails to honestly address the cost of Yucca Mountain, and fails to provide a workable funding mechanism for the restarted repository program. The Subcommittee's consideration of H.R. 2699 should begin by reviewing repository costs, starting with the DOE 2008 Total System Life Cycle Cost (TSLCC) Analysis and the 2013 DOE Fee Adequacy Report. We estimate the total future cost of Yucca Mountain would be at least $100 billion in 2019 dollars. Licensing alone would cost $2 billion over 4 or 5 years. DOE studies prepared between 2010 and 2013 estimated that walking away from Yucca Mountain and constructing a repository in salt or shale could save tens of billions of dollars. The Committees on Appropriations in both houses should require an updated estimate of Yucca Mountain costs, and the estimated costs of constructing repositories in other rock types, with alternative repository designs, before appropriating any new funds for Yucca Mountain licensing.

Stranded Spent Nuclear Fuel at Decommissioned and Decommissioning Reactors

The Subcommittee is considering two bills today that would address the adverse impacts on host communities of stranded spent nuclear fuel at decommissioned and decommissioning reactors.

H.R. 2995, the Spent Fuel Prioritization Act of 2019, and H.R. 3136, the Storage and Transportation of Residual and Excess (STORE) Nuclear Fuel Act of 2019 are both intended to address the growing safety and economic concerns of impacted communities. We have not yet had sufficient time to fully evaluate Section 3, Limitation on Collection of Fees and Section 4, Funding in H.R. 3136. But, our preliminary finding is that these two bills, taken together, would be an important step towards implementing the 2012 Recommendations of the Blue Ribbon Commission on America's Nuclear Future, and the 2018 recommendations of the Western Interstate Energy Board, that removal of spent nuclear fuel from shutdown reactor sites be prioritized.

H.R. 3136 would be more effective than H.R. 2699 in expediting removal of stranded fuel from decommissioned and decommissioning sites. Under H.R. 2699, the Monitored Retrievable Storage (MRS) facility could not receive spent fuel before a final NRC licensing decision approving or disapproving Yucca Mountain. H.R. 2699 retains the 10,000 MTHM capacity limit on MRS spent fuel storage until the repository first accepts spent fuel, and limits MRS capacity to 15,000 MTHM at all times. These MRS conditions would severely limit acceptance of spent fuel from currently decommissioned or decommissioning reactors, and could make MRS development unattractive to private companies.

H.R. 3136, combined with H.R. 2995, would be more effective in expediting removal of stranded fuel than resuming the Yucca Mountain licensing proceeding. The earliest realistic date that the proposed Yucca Mountain repository could receive spent fuel from decommissioned and decommissioning reactors would be 20-25 years from now, if ever. Nevada never stopped preparing for the resumption of licensing. Under the leadership of the Nevada Office of Attorney General, Nevada's legal and technical expert team has been in place since 2010 preparing to adjudicate 218 admitted contentions and to submit 30-50 new contentions using state funds appropriated by the Nevada Legislature. Any effort to speed up the process by imposing time limits on licensing, as proposed in H.R. 2699, would almost certainly backfire, increasing costs, increasing safety disputes, and increasing litigation.

An additional consideration is that the spent fuel at shutdown and soon-to-be-retired reactors is being welded into storage canisters that are not compatible with DOE's Yucca Mountain repository design. The spent fuel inside the canisters currently used for dry storage at U.S. reactor sites would need to be repackaged, or the Yucca Mountain waste package design would need to be changed, or a combination of both actions would be necessary, increasing by years the lead time for acceptance of U.S. reactor spent fuel at Yucca Mountain.

Consent-Based Siting for All Nuclear Waste Storage and Disposal Facilities

The State of Nevada supports H.R. 1544, the Nuclear Waste Informed Consent Act, introduced in March 2019 by Representatives Titus, Horsford, and Lee of Nevada. H.R. 1544 would require a written consent agreement between

DOE, the repository host state, affected counties, and affected Indian Tribes, prior to construction of a repository. This would extend consent to the State of Nevada for Yucca Mountain. A companion bill, S. 649, was introduced by Senators Catherine Cortez Masto and Jacky Rosen of Nevada. H.R. 1544 and S. 649 provide a basis for amending other bills to create a workable approach to consent-based siting for all U.S. nuclear waste storage and disposal.

The Nevada Commission on Nuclear Projects recommended this approach in 2017: "In the past two Congresses, the Senate Energy and Natural Resources Committee has drafted comprehensive legislation, entitled the Nuclear Waste Administration Act, to restructure the nation's nuclear waste program following the BRC recommendations. This legislation is not acceptable to the State of Nevada because it would continue the status quo regarding Yucca Mountain. It would need to be amended along the lines of the Nuclear Waste Informed Consent Act, introduced by the Nevada congressional delegation."[7]

The Nuclear Waste Administration Act of 2019, S. 1234, would create a new waste management organization called the Nuclear Waste Administration (NWA); direct the NWA to establish a consent-based siting process; and calls for operation of a spent nuclear fuel storage pilot facility by December 31, 2025, an interim storage facility for spent nuclear fuel by December 31, 2029, and a geologic repository by December 31, 2052 [page 64, lines 19-24]. These storage and disposal facilities would be regulated by the NRC, subject to standards established by the EPA S. 1234 could be amended to extend the new consent-based siting process to Nevada regarding Yucca Mountain.

Governor Steve Sisolak concluded his letter with a pledge: "If your committee is truly interested in fixing our nation's broken nuclear waste program, my staff and I, and Nevada's congressional delegation, would be happy to meet with you and explore constructive alternatives." I hope the Subcommittee and the full Committee will consider my testimony today as a first step in fulfilling Nevada's part of the Governor's pledge.

* * *

Mr. TONKO. The gentleman concludes his statement. We thank Mr. Halstead.

Next, we recognize Mr. Keyser for five minutes, please.

---

[7] http://www.state.nv.us/nucwaste/news2017/pdf/nv2017comm_report_final.pdf [p.27].

Statement of Austin Keyser

Mr. KEYSER. Chairman Tonko, Ranking Member Shimkus, and members of the House Energy and Commerce Subcommittee on Environment and Climate Change, thank you for inviting me.

My name is Austin Keyser. I am the director of the International Brotherhood and Electrical Workers Political and Legislative Affairs Department. I have been asked by our president, Lonnie Stephenson, to speak on his behalf.

The IBEW is the largest energy union in the world. We represent more than 775,000 members in the United States and Canada, and we work in a variety of energy related fields including utilities, construction, telecommunications, broadcasting, manufacturing, railroads, and government.

Nuclear energy in America produces 20 percent of our electricity and accounts for 50 percent of all zero-carbon power. The need for the country's only carbon-free-source that can ensure around-the-clock generation had become even greater as we move toward more renewable energy.

Supporting nuclear generation is critical if the U.S. is going to reduce emissions while meeting our increasing base load energy demands. Fifteen thousand members of the IBEW are employed fulltime by the nuclear industry at 55 facilities across the United States. Thousands more IBEW members rotate through nuclear plants for maintenance and refueling. The IBEW's history in the industry coincides with the construction of the very first nuclear facilities.

The nuclear industry supplies high wages and safe jobs. These options pay one-third more than the average jobs in the communities where they exist and tout safety records that are the high watermark for American industry. These are the types of family-sustaining careers that Americans are looking for and policymakers should support. Hundreds of IBEW members lose their jobs every time a nuclear facility closes often eliminating the biggest source of economic activity in the community. We have already seen the adverse impacts that plant disclosures cause. We fear similar fate for workers at Davis-Besse and Perry in my home State of Ohio and other communities at risk of losing their facilities.

But critical to the future of the Nation's nuclear sector is opening a permanent repository for spent nuclear fuel, SNF. Under Federal statute, a permanent geologic repository was scheduled to open at Yucca Mountain Nevada no later than January 31, 1998. Over two decades later, ratepayers and workers are still waiting for a place to safely store over 80,000 metric tons of SNF sitting at 121 sites in 35 States across the country.

The IBEW supports effort to restart the licensing process of Yucca Mountain. Going back to the late 1970s, the IBEW has endorsed legislation that ensures timely central storage, safe transportation, and permanent disposal of spent nuclear fuels. The IBEW was part of the construction of Yucca Mountain site before work halted in 2010. If work recommences there, it would create 2,000 construction jobs lasting a decade. A permanent repository would help alleviate concerns about nuclear power and boost support for nuclear generation as a foundational part of Nation's energy portfolio.

A permanent repository is also necessary to ensure public support for the next generation of advanced nuclear reactors that we hope will come online in the near future. Additionally, the IBEW supports the authorization of a consolidated interim facility to safely store SNF particularly for dry caskets and nongenerating nuclear plants.

An interim facility would allow for the redevelopment of shuttered nuclear plants bringing economic revitalization, tax review and jobs to working families and communities that were hard hit by the closures. Many closed stations are ideal sites for future development of other forms of electrical generation, including renewables, due to the existing electrical transmission infrastructure. A consolidated interim storage facility should have the support of the host State and local community. Authorizing legislation should have clear language ensuring that an interim facility does not become a de facto permanent storage site.

Today's hearings focus on three bills. The Nuclear Waste Policy Amendments Act, the Storage and Transportation Residual and Excess Act, and the Spent Fuel Prioritization Act. The IBEW supports the opening of a permanent storage facility as soon as one is properly licensed to build. H.R. 2699 would help resolve key issues such as land withdrawal and site infrastructure at Yucca Mountain and require the NRC to conclude its review of the application within 30 months of enactment.

This legislation would authorize the Energy Department to open interim storage facilities to consolidate SNF from sites with a decommissioned and provide a pathway for the State of Nevada and local communities to discuss benefits associated with these projects.

Similar legislation passed the House of Representatives last year with strong bipartisan support. H.R. 3136 would authorize the Energy Department to open at least one interim facility. It would require consent from the State and local community before an interim site could be licensed.

We support the provisions in the STORE Act to prioritize SNF at closed nuclear facilities for interim storage. The IBEW would strongly prefer that Congress take action to open a permanent repository as soon as possible. But we recognize that an interim facility may be the best first step toward a comprehensive solution that will consolidate SNF at a central repository. H.R. 2995 would prioritize storage of SNF from decommissioned nuclear plants near large populations and seismic hazards.

Congress and Federal regulators should treat all communities fairly by also considering the length of time the plant has been decommissioned and the other potential environmental risks. The IBEW supports congressional action that will protect ratepayers who have paid tens of billions of dollars into the Nuclear Waste Fund for decades by ensuring fees are not prematurely reinstituted until it is demonstrated that additional moneys are necessary to achieve full cost recovery of the program.

We appreciate that the sponsors of the Nuclear Waste Policy Amendments Act and the STORE Act included these protections.

In conclusion, the IBEW respectfully urges Congress to take the necessary steps to open a permanent repository. Now is the time to come together and pass bipartisan legislation that will honor the Federal Government's promise to the nuclear industry, union workers, and ratepayers.

Thank you.

[The prepared statement of Mr. Keyser follows:]

Testimony of Austin Keyser, Director, Political & Legislative Affairs, International Brotherhood of Electrical Workers, before the Subcommittee on Environment and Climate Change, House Committee on Energy and Commerce, U.S. House of Representatives, June 13, 2019

Chairman Tonko, Ranking Member Shimkus, and Members of the House Energy and Commerce Subcommittee on Environment and Climate Change, thank you for inviting me to today's legislative hearing.

Background

My name is Austin Keyser. I am the Director of the International Brotherhood of Electrical Workers Political & Legislative Department. I have been asked by our President, Lonnie Stephenson, to speak on behalf of the IBEW.

The IBEW is the largest energy union in the world. We represent more than 775,000 members in the United States, U.S. territories, and Canada, who work in a variety of energy related fields including utilities, construction, telecommunications, broadcasting, manufacturing, railroads, and government.

Importance of Nuclear Power

Nuclear energy in America produces nearly 20 percent of our nation's electricity and accounts for over 50 percent of all zero carbon emission generation in the country. As the United States moves towards increasing reliance on renewable energy, such as solar and wind, the need for nuclear energy's reliability, the country's only carbon-free source that can ensure around-the-clock generation, even during inclement weather, has become even greater. Supporting nuclear generation is critical if the United States is going to continue to reduce carbon emissions and avoid the worst potential impacts of climate change while meeting baseload energy demands for American citizens and businesses.

Nuclear Power and the Workforce

Fifteen thousand members of the IBEW are employed full-time by the nuclear industry at over 55 facilities across the United States. Thousands more IBEW members rotate through nuclear plants with the contractor workforce as needed for maintenance and refueling. The IBEW has a proud history of working in the commercial nuclear industry dating back to the 1950s with the first civilian nuclear reactor at Shippingport, Pennsylvania. IBEW nuclear workers can say without reservation that this is an industry with a proven record of exceptional safety. It is among the safest industrial work environments in the United States.

The nuclear industry supplies high quality, long term, steady work that typically pays one-third more than the average jobs in their community. These are the types of family-sustaining careers that Americans are looking for and policymakers should support.

A major issue of concern for the IBEW's membership are the large number of nuclear plant closures that have taken place in recent years and the additional closures that are anticipated in the near future. Hundreds ofIBEW members lose their well-paying, family-supporting jobs every time a nuclear facility closes, often eliminating the biggest source of economic activity in their respective communities. We've already seen the impacts of plant closures for communities that supported Vermont Yankee, Fort Calhoun and Kewaunee. We fear a similar fate for our workers and their communities, including for IBEW workers at the Davis-Besse and Perry plants in my home state of Ohio, for the Beaver Valley and Three Mile Island facilities in Pennsylvania, and for workers at the recently decommissioned Pilgrim nuclear plant in Massachusetts.

Spent Nuclear Fuel

A critical piece to supporting the future of our nation's nuclear sector, and the tens of thousands of family-supporting jobs that the nuclear industry creates, is opening a permanent repository for spent nuclear fuel (SNF). Under federal statute, a permanent geologic repository was scheduled to open at Yucca Mountain, Nevada no later than January 31, 1998. Over two decades later, ratepayers and workers are still waiting for a repository to safely store over 80,000 metric tons of SNF sitting at 121 sites in 39 states across the country. The IBEW supports legislative efforts to restart the licensing process of Yucca Mountain and allow the Nuclear Regulatory Commission (NRC) to make a final decision on whether the site meets the necessary requirements to store our nation's SNF. Going back to the late 1970's, the IBEW has adopted formal resolutions that endorse legislation that "ensures timely central storage, safe transportation, and permanent disposal of spent nuclear fuels."

The IBEW and the other members of the Building Trades were part of the construction of the Yucca Mountain site before work halted in 2010. If work recommences at Yucca Mountain, it is estimated it would create 2,000 construction jobs for 10 years.

The opening of a permanent geologic repository would help alleviate the concerns within the general public surrounding nuclear power and boost support for nuclear generation as a foundational part of our nation's energy portfolio. We believe a permanent repository is necessary to ensure the public's support for the next generation of advanced nuclear reactors that we hope will come online in the near future, including small modular reactors (SMRs).

Additionally, the IBEW supports the authorization of a consolidated interim facility to safely store SNF, particularly for dry caskets currently located at non- generating nuclear plants. The opening of an interim storage facility would allow for the redevelopment of shuttered nuclear plants, bringing economic revitalization, tax revenue and jobs to working families and communities that were hard hit by these closures. In fact, many closed nuclear stations are ideal sites for future development of other forms of electrical generation, including renewables, due to the already existing electrical transmission infrastructure.

It is important that the opening of a consolidated interim storage facility have the support of the host state and local community and the authorizing legislation have clear language ensuring that an interim facility does not become a de facto permanent storage site.

Legislation

Today's hearing focuses on three bills: the Nuclear Waste Policy Amendments Act (H.R. 2699), the Storage and Transportation of Residual and Excess (STORE) Act (H.R. 3136), and the Spent Fuel Prioritization Act (H.R. 2995).

The IBEW, as stated earlier, supports the opening of a pennanent geologic storage facility as soon as one is properly licensed and built. The Nuclear Waste Policy Amendments Act would help facilitate the licensing process by resolving key issues such as land withdrawal and site infrastructure at Yucca Mountain and require the Nuclear Regulatory Commission to conclude its review of the application within 30 months of enactment. This legislation, in addition, would authorize the Energy Department to open interim storage facilities to consolidate SNF from sites with a decommissioned reactor and provide a pathway for the State of Nevada and local communities to discuss benefits associated with these projects. Very similar legislation passed the House of Representatives last year with strong bipartisan support and a 340-72 vote.

The STORE Act would authorize the Energy Department to open at least one interim storage facility. The legislation, similarly to H.R. 2699, would require consent from the state and local community before an interim site can be licensed. Additionally, we support the provisions in the STORE Act to prioritize SNF currently at closed nuclear facilities for interim storage.

The IBEW would strongly prefer that Congress take action to open a permanent repository as soon as possible, but given the current political barriers surrounding Yucca Mountain, we recognize that providing authorization for an interim facility may be the best first step towards a necessary comprehensive solution that will consolidate SNF at a central repository.

The Spent Fuel Prioritization Act would prioritize the storage of SNF from decommissioned nuclear plants near large populations and located in an area with a high seismic hazard. The IBEW believe Congress and federal regulators should treat all communities with decommissioned nuclear facilities fairly and consider several factors to best prioritize storage, including the length of time the plant has been decommissioned and potential harm to the public, including from decaying containers, seismic activity and rising sea levels.

The IBEW, in addition, supports congressional action that will protect ratepayers, including union workers, who have paid tens of billions of dollars into the Nuclear Waste Fund for decades by ensuring fees are not prematurely reinstituted until it is demonstrated that additional monies are necessary to achieve full cost recovery of the program. We appreciate that the sponsors of the Nuclear Waste Policy Amendments Act and the STORE Act included provisions to protect ratepayers from such additional fees.

Conclusion

On behalf of the IBEW, I again thank the committee for the opportunity to testify this morning and thank the sponsors of the three bills that are being discussed. The IBEW respectfully urges Congress to take the necessary steps to resolve this problem and open a permanent repository. Now is the time to come together and pass bipartisan legislation that will honor the federal government's promise to the nuclear industry, union workers and ratepayers.

* * *

Mr. TONKO. Thank you, Mr. Keyser.

And, Mr. Fettus, you are recognized for five minutes, please.

Statement of Geoffrey H. Fettus

Mr. FETTUS. Thank you, Chairman Tonko, Ranking Member Shimkus, and members of the subcommittee. Thank you for the opportunity to present NRDC's views. We hope this hearing can be a new beginning. With more than 80,000 metric tons in more than half of our States and reactors moving to decommissioning, we need to reset the process. The drafts before us today, however well intentioned, will not solve the current stalemate and won't lead toward workable solutions.

For more than 50 years, Congress has offered and sometimes passed bills that are variations on these themes. Restart the Yucca licensing process or kick open a door in New Mexico for an interim storage site. When that State was promised repeatedly, no such thing would ever happen.

We sit here today because these efforts have failed repeatedly over decades. In Tennessee, in Kansas, Nevada Utah, everywhere else. Another such attempt restarts litigation and controversy, the likely result if these bills move forward: continued stalemate.

Seven years ago, President Obama's bipartisan blue ribbon commission keenly described why past attempts failed. That Presidential commission wisely asserted we can't keep doing the same thing, which is what we are doing here today. Congress must create a process that allows any potential host State to consent and, for that matter, not consent.

Our written testimony addresses why these specific drafts won't work. So, with these four minutes, I present for your consideration a durable reset of how we can manage and dispose of nuclear waste and how you can and should take the lead.

Our solution can be summed up simply: Give EPA and the States power under our bedrock environmental statutes so that the States can set the terms for how much and on what conditions they could host disposal sites.

How would reapportioning the power change things? I urge you to look at the root of why we are stuck in half of a century of rancor. Radioactive waste is stranded because the Atomic Energy ACT treats it as a privileged pollutant. The act pre-empts EPA and State regulatory authority, exempting radioactivity from hazardous waste law, sizeable portions of the Clean Water Act, and several others.

By disconnecting radioactivity from the normal patterns of environmental law, we ignore the vital role States play in addressing environmental contaminates, protecting their citizens, and generally regulating what happens within their borders. We can open a path forward that respects each State rather than offering up the latest one for sacrifice. With the protection of the environmental statutes, a State can say no or yes and on what terms and not necessarily be subject to hosting the entire burden.

Such a new regime would allow for a thorough technical review without the fear that negative findings will be dispensed with as soon as it is expedient, as happened repeatedly with the Yucca process. Deep scientific review could be at the forefront, this time joined with institutional structures that would allow for public acceptance of solutions.

Our government is at its strongest when each player's role is respected. State consent and public acceptance of potential repository sites will never be willingly granted—and we have 50 years of evidence for that, including from New Mexico, and I am hearing Texas right now—unless and until power on how, when, and where the waste will be disposed of is shared rather than decided by Federal fiat.

I am sorry I can't give you a guarantee that eliminating these exemptions will magically solve the puzzle, but I can guarantee that trying to force open a storage or disposal site over the objections of a State, whether it is Utah, New Mexico, Texas, or Nevada will only lead to more of the same, and then the waste won't move.

We have seen these bills before, but each has been a mirror of the last. NRDC is not saying no but how to get to yes and how to do so in a way that could work in our democracy. Strong environmental laws have a proven track record. Giving States that regulatory authority will give them the ability to consent on scientific and political terms that they can live with. It is time to regulate nuclear waste the same way as every other pollutant with EPA and the delegated States taking the lead under our foundational environmental statutes.

Thank you again for having me here, and I look forward to answering your questions.

[The prepared statement of Mr. Fettus follows:]

Statement of Geoffrey H. Fettus, Senior Attorney, Natural Resources Defense Council, NRDC Statement for Legislative, Hearing on "Cleaning up Communities: Ensuring Safe Storage and Disposal of Spent Nuclear Fuel," before the Congress of the United States, United States House of Representatives, Committee on Energy & Commerce, Subcommittee on Environment and Climate Change, Room 2322, Rayburn House Office Building, June 13, 2019

Introduction & Summary

Mr. Chairman, Mr. Ranking Member and members of the Subcommittee, thank you for providing the Natural Resources Defense Council, Inc. (NRDC) this opportunity to present our views on H.R. 2699, "Nuclear Waste Policy Amendments Act of 2019"; H.R. 3136, "Storage and Transportation Of Residual and Excess Nuclear Fuel Act of 2019"; and H.R. 2995, "Spent Fuel Prioritization Act of 2019." We appreciate that the Committee sees the need to commence work again on solving our national nuclear waste dilemma and we hope to work with all of you on a constructive process.

NRDC is a national, non-profit organization of scientists, lawyers, and environmental specialists, dedicated to protecting public health and the environment. Founded in 1970, NRDC serves more than three million members, supporters and environmental activists with offices in New York; Washington, D.C.; Los Angeles; San Francisco; Chicago; Bozeman, Montana; and Beijing. We have worked on nuclear waste matters since our founding and continue to do so.

We are cognizant, as are all of you, of the veritable tsunami of legislative history detailing objections or support to similar pieces of legislation as those before us today. Indeed, we've contributed to that wealth of testimony and did so before this Committee only a few years ago.[8] With that in mind, we will use this valuable time to summarize our position on key issues in each draft

---

[8] *See, e.g.*, Hearing Before The Subcommittee On Environment Of The Committee On Energy And Commerce, House Of Representatives, 115th Congress, First Session April 26, 2017, Serial No. 115–26, online at https://www.govinfo.gov/content/pkg/CHRG-115hhrg25996/pdf/CHRG-115hhrg25996.pdf; *see* NRDC's submission at 136 (pdf page 140 of 172); or NRDC's 2015 Testimony Before the House Energy & Commerce Committee; *see* https://www.nrdc.org/experts/matthew-mckinzie/nrdc-testifies-house-representatives-nuclear-waste.

and then turn to what we hope is the productive step of charting an alternative, pragmatic legislative path forward.

Each item under discussion today is premised on a good intention – finding a way forward on storing or disposing of commercial spent nuclear fuel. There are nods to garnering consent; there are efforts to create a sensible process for shipping nuclear waste when the day finally comes for transport away from reactors; and there are overdue acknowledgements that communities that have benefitted from nuclear power, but also bear the burden of hosting the waste, need support. But even with these constructive thoughts, if these drafts were enacted, they would set us back and could derail our efforts at scientifically defensible and publicly accepted solutions to the nuclear waste problem in the United States. Thus, we testify today in opposition to moving forward on these drafts and suggest another course.

Here's why. Enacting what is on offer today would immediately precipitate a welter of controversy and litigation from the potential recipient states, which will result in no progress toward a solution and years more rancor. Witness, as a keen example, the Private Fuel Storage interim nuclear waste storage site in Utah, which was licensed in 2006 but has not, nor will foreseeably receive waste due to the state's steadfast resistance. Enacting these drafts would also continue all the attendant frustrations that come with nuclear waste in pools or dry storage at Nuclear Regulatory Commission (NRC) licensed reactors around the country. Seven years ago, President Obama's Blue Ribbon Commission on America's Nuclear Future rightly found that consent-based siting, with meaningful partnerships and open communication among federal, state, local, and tribal leaders, is the most important step toward establishing geologic nuclear waste repositories.[9] These drafts bypass that wise observation, and try a slight variation of the same approach of forcing the waste on Nevada, New Mexico, and Texas (or elsewhere).

There is another way forward; one that defuses the rancor before it begins. A legislative change that would provide potential host states the right to say "No," but also "Yes, and on these strict, protective terms, and with these distinct limits." A legislative change that might not address all the waste at once, but it can get us started and likely in a much faster time frame than attempting to fight Nevada once again. This path forward can happen if Congress fixes the fundamental flaw in the Nuclear Waste Policy Act, 42

[9] President Obama's "*Blue Ribbon Commission on America's Nuclear Future - Report to the Secretary of Energy, January 31, 2012*" (hereafter "*BRC*"); *see* online at https://www.energy.gov/sites/prod/files/2013/04/f0/brc_finalreport_jan2012.pdf.

U.S.C. §10101 *et seq*. (NWPA) – the exemption of radioactivity from environmental laws that has been part and parcel of the Atomic Energy Act for decades. Ending this exemption through legislative change will then allow for meaningful, full regulatory authority from the EPA and the potential host states. These bills won't get us to moving waste off reactor sites, but this change in law can.

NRDC's Analysis of the Draft Legislation Presented

While all three draft bills begin with good intentions and have elements of constructive policy, they don't collectively, or on their own, chart a way forward. We briefly discuss some key points from each in turn.

H.R. 2699, "Nuclear Waste Policy Amendments Act of 2019"

H.R. 2699 is the reprise of an effort we've seen over the last several years, with a few changes. Our reactions in 2015 and 2017 (see citations, n. 1), remain the same. The provisions we briefly address follow in italics, with our observations in regular text.

> Title 1, Sec. 101 (a)(4) The Secretary shall, not later than 90 days after the date of enactment of the Nuclear Waste Policy Amendments Act of 2019, publish a request for information to help the Secretary evaluate options for the Secretary to enter into cooperative agreements with respect to one or more monitored retrievable storage facilities.

Respectfully, this provision, and the other provisions in Title 1, and Title III (where DOE can take title to all the nuclear waste) are premature. First, immediately going forward with a consolidated storage proposal (which is equivalent to a "monitored retrievable storage facility") before working out the details of a comprehensive legislative path to solve the nuclear waste problem (firmly and irrevocably connecting the licensing of storage to the licensing of a permanent repository) severs the link between storage and disposal and creates an overwhelming risk that an interim storage site will become a de facto final resting place for nuclear waste. Or, in the alternative and also just as damning, it sets up yet another attempt to ship the waste to Yucca Mountain irrespective of regulatory process, or open up New Mexico's Waste Isolation Pilot Plant (WIPP) facility for spent nuclear fuel disposal – the latter site designed and intended for defense-generated nuclear waste with trace levels of plutonium, not spent fuel (and NRDC notes, a site that has already seen an accident dispersing plutonium throughout the underground and into the environment, contaminating 22 workers and making the site functionally inoperable for years). All of this runs precisely counter to the core admonition of President Obama's Blue Ribbon Commission (BRC) that "consent" come first.

We appreciate the intent of Title I, section 103's expansion of NWPA section 143(b) from previous iterations of the bill, as it makes a nod toward keeping some fig leaf of a link between storage and disposal. But once a site has been forced to start accepting waste, any future Congress can (and we have strong historical evidence they likely will) simply override any single state's objection to altering the terms of how much waste or on what timeline waste will be sent once that site has become a de facto repository.

> Sec. 102(b) AUTHORIZATION ... the Secretary is authorized to—(1) site, construct, and operate one or more monitored retrievable storage facilities; and (2) store, pursuant to a cooperative agreement, Department-owned civilian waste at a monitored retrievable storage facility for which a non-Federal entity holds a license described in section 143(1).
>
> (c) PRIORITY.—(1) IN GENERAL.—Except as provided in paragraph (2), the Secretary shall prioritize storage of Department-owned civilian waste at a monitored retrievable storage facility authorized under subsection (b)(2). (2) EXCEPTION.—(A) DETERMINATION.—Paragraph (1) shall not apply if the Secretary determines that it will be faster and less expensive to site, construct, and operate a facility authorized under subsection (b)(1), in comparison to a facility authorized under subsection (b)(2).

In this provision, regrettably there are no safety, environmental or public acceptance criteria cited, only the expediencies of speed and reduction of expense. This is precisely the formula that produced the failure of the Yucca Mountain process and made it, as the previous administration noted, "unworkable." Congress should turn away from this path and, rather, pursue a phased, adaptive, consent-based and science-based siting process as the best approach to gain the public trust and confidence needed to site nuclear waste facilities. Trust most of all is what is missing from the nuclear waste predicament – trust that what is promised in the present will be not altered for others' convenience in the future. As we describe later in detail, the solution to regaining trust is that Congress must finally end the Atomic Energy Act (AEA) exemptions from environmental laws. By providing our hazardous waste and clean water laws full authority over radioactivity and nuclear waste facilities so that EPA and – most importantly – the states can assert direct regulatory authority, Congress can provide EPA and the states with familiar authority in which they can place ongoing trust.

Title II, Sec. 201 Land Withdrawal, Jurisdiction, and Reservation

These provisions, of which we only note the title, run counter to state authority and likely lead to the same kind of stalemate that currently exists this day. NRDC expects that Nevada can and will ferociously object to this usurpation of its rights. Indeed, this kind of legislative alteration of rights sets the stage for other states to beware if and when nuclear waste projects commence in their states (see, *e.g.*, New Mexico, and the changes necessary if Congress were to decide WIPP might provide a final disposal option).

> Sec. 202 (b)(3) At any time before or after the Commission issues a final decision approving or disapproving the issuance of a construction authorization for a repository pursuant to paragraph (1), the Secretary may undertake infrastructure activities that the Secretary considers necessary or appropriate to support construction or operation of a repository at the Yucca Mountain site or transportation to such site of spent nuclear fuel and high-level radioactive waste. Infrastructure activities include safety upgrades, site preparation, the construction of a rail line to connect the Yucca Mountain site with the national rail network (including any facilities to facilitate rail operations), and construction, upgrade, acquisition, or operation of electrical grids or facilities, other utilities, communication facilities, access roads, and nonnuclear support facilities.

This is premature and wasteful spending of taxpayer dollars on Yucca Mountain infrastructure prior to meaningfully restarting the licensing process, which we think would be fruitless. Only a few years ago, the NRC put the cost of completing the Yucca Mountain license application at more than $330 million. Efforts to revive this process would be at best problematic and likely waylay the process of developing a repository for a significant period of time. Briefly, there are hundreds of license contentions to be litigated at significant length and cost. One contention in particular contests the viability of the Department of Energy's (DOE's) design for titanium drip shields that are intended to be placed on top of each of the thousands of waste canisters in Yucca Mountain's underground tunnels to deflect corroding water. Although DOE included the drip shields as part of the repository design, and NRC has accepted them for license-review purposes, there is no plan to engineer, license, pay for, and much less install the shields until at least 100 years after the waste goes in. Quite simply, it seems fair to suggest Yucca's likely repository configuration will not meet NRC requirements.

This and other issues will be vigorously litigated by Nevada, which has filed more than 200 contentions challenging DOE's license application for Yucca Mountain. To put such a hearing process in perspective, NRDC concluded nearly six years of a NRC licensing proceeding where not one party – not industry seeking the license, not NRC Staff, nor the environmental intervenors – had any interest or took any steps to functionally prolong or delay the proceeding. And in the more than five years of this proceeding, only three contentions were fully litigated on their merits, not the more than 200 teed up for the Yucca hearing if it were to ever be resumed. Any suggestion that the Yucca licensing proceeding could easily restart and quickly move to a successful conclusion for permanent disposal is a fallacy. And when that inevitable litigation rightly waylays yet another effort at a long-ago vetoed nuclear waste disposal proposal, the damage to the nation's prospects to ever developing a repository may be long-lasting. Rather than go forward with construction as this provision allows, we urge Congress to take the careful course of building a publicly accepted, consent-based repository program.

> Sec. 202 (b)(3)(B) & (C) If the Secretary determines that an environmental impact statement is required under the National Environmental Policy Act of 1969 with respect to an infrastructure activity undertaken under this paragraph, the Secretary need not consider the need for the action, alternative actions, or a no-action alternative ... (C) NO GROUNDS FOR DISAPPROVAL.— The Commission may not disapprove, on the grounds that the Secretary undertook an infrastructure activity under this paragraph— (i) the issuance of a construction authorization for a repository pursuant to paragraph (1); (ii) a license to receive and possess spent nuclear fuel and high-level radioactive waste; or (iii) any other action concerning the repository.

These provisions conflict with well-established and necessary requirements of the National Environmental Policy Act, 42 U.S.C. §4321, *et seq.* (NEPA) and make it clear that compliance with longstanding environmental laws will be an afterthought. Doing so exacerbates the public interest community's (and that of Nevada) objection of the last three decades that the process of developing, licensing, and setting environmental and oversight standards for the proposed repository has been, and continues to be, rigged or weakened to ensure that the site can be licensed, rather than provide for safety over the length of time that the waste remains dangerous to public health and the environment.

> Sec. 205 It is the Sense of Congress that the Secretary of Energy should consider routes for the transportation of spent nuclear fuel or high-level radioactive waste transported by or for the Secretary under subtitle A of title I of the Nuclear Waste Policy Act of 1982 (42 U.S.C. 10131 et seq.) to the Yucca Mountain site that, to the extent practicable, avoid Las Vegas, Nevada.

This provision illustrates the difficulties faced in trying to force a project on an unwilling and non-consenting host state. Are other cities along the routes in other states that will be affected by the avoidance of Las Vegas allowed to have a say in the matter? And if the NEPA process is truncated, is there any meaningful opportunity to have that say? Lack of consent from an unwilling host state is a recipe for disaster, whether the issue is nuclear waste or any other great public concern. We include a map (NRDC Att. 1) that highlights the vast number of routes through many states that will be taken by nuclear waste on its way to Yucca Mountain. One can visualize the extraordinary number of regions and people that will be affected by this enterprise, and without a very different process to arrive at public acceptance of the process of developing those routes and the associated burdens, we see a continuation of the stalemate.

> Sec. 601(a)(1)(A) Not later than 2 years after the Nuclear Regulatory Commission has issued a final decision approving or disapproving the issuance of a construction authorization for a repository under section 114(d)(1) of the Nuclear Waste Policy Act of 1982 (42 U.S.C. 10134(d)) (as so designated by this Act), the Administrator of the Environmental Protection Agency shall (A) determine if the standards promulgated under section 121(a) of the Nuclear Waste Policy Act of 1982 (42 U.S.C. 10141(a)) should be updated; and (B) submit to Congress a report on such determination.

This provision has the matter precisely backwards. To avoid repeating failures of past decades and consistent with BRC recommendations, both the standards for site screening and development criteria must be in final form before any sites are considered. Generic radiation and environmental protection standards must also be established prior to consideration of sites. We expand upon this later (*infra*, 13, 14).

H.R. 3136, "Storage and Transportation of Residual and Excess Nuclear Fuel Act of 2019"

We noted at the outset that H.R. 3136 is premised on the good intention to get the process moving, but if enacted, this would surely precipitate a welter of litigation. And if any nuclear waste storage facility were successfully constructed under the authority this law would provide, there are no limits in place (including neither quantity of waste or timing of licensing) preventing such a facility from becoming a de facto permanent site, as the law completely severs the link between storage and repository.

Further, while we appreciate the nod to consent in Sec. 191(f)(4), as New Mexico's recent objection letter from Governor Lujan Grisham (NRDC Att. 2) makes plain, there is no simple path forward. Requiring consent to site a facility does not provide a state, unit of local government, or tribe any recourse against future bad acts once the facility is approved. And removing the ability of the United States to unilaterally break the terms of the contract (under the "Binding Effect" provision in subsection B of the Section), could potentially give a state an artificial measure of political comfort that the agreement it had painstakingly negotiated will hold fast. But there would be nothing stopping Congress from revisiting this law, ratifying the consent agreements with conditions, and thereby removing whatever meaningful restraint a state might assert. Thus, ultimately what is offered as a thoughtful contract provision could be rendered inoperable, and could eviscerate a state's protection against altered, less favorable terms. Bluntly, efforts to force the interim storage process forward prior to putting together a scientifically defensible and publicly accepted program that provides states with durable authority is unwise and likely to derail progress yet again.

Rather than prematurely bypassing a careful, consent-based process, NRDC urges Congress to require the NRC and industry to focus spent fuel storage efforts on ensuring that all near-term forms of storage meet high standards of safety and security for the decades-long time periods that interim storage sites will be in use.

H.R. 2995, "Spent Fuel Prioritization Act of 2019"

We have no quarrel with the idea of a health and safety prioritization process for the queue of spent nuclear fuel that will, someday, go deep underground in a geologic repository. It's a constructive, thoughtful part of the discussion that needs a thorough airing to make wise policy. But not right now, for two key reasons.

First, reordering the queue of waste for a repository program that has been all but defunct for years puts the cart before the horse. We don't even have a path forward at present. Neither Yucca Mountain nor the current interim storage ideas are likely to have success, and suggesting a reordering of the waste that may not move for decades if the impasse continues is misplaced.

Rather, Congress directing NRC to require improved storage at the reactor site – such as legislatively directing the NRC to require much quicker movement to hardened onsite storage and not allowing for long term storage at inappropriate locations, like a beach in California, when other, better options at the utility/reactor controlled land exist – is a much more sensible approach for the near term while Congress works out a better legislative pathway for durable solutions.

Second, any reordering of the nuclear waste queue has the potential for a change in the ongoing burden on a huge number of host communities and regions. While all of these communities benefited from the power generation, they now bear the burden of hosting the dangerous nuclear waste, and all that is entailed with that, for decades to come. This is something that needs to be done carefully, and the consent/regulatory authority we will describe in the coming pages can provide a suitable pathway for this dialogue to happen.

Understanding the History & Need for a Fundamental Change in Law

As detailed above, H.R. 2699 and H.R. 3136 attempt to clear the legal obstacles to allow New Mexico or Texas to receive sizable portions of the nation's nuclear waste at a consolidated interim storage site that has significant legal and technical challenges and is opposed by the entire New Mexican Congressional delegation and Governor.[10] Title II of H.R. 2699 sets the abandoned, defunct Yucca Mountain licensing process back in motion, but with an even more truncated environmental review, and with a set of new potential sources of state funding. Nevada issued its notice of disapproval of Yucca Mountain on April 8, 2002 and has repeatedly stated its opposition, seemingly to no avail. Respectfully, none of this will work. There is a better way.

---

[10] See NRDC Att. 2. Also, we note that while there is a nod toward "consent" in the text of the legislation, *see* Section 143(a)(2), Conditions for MRS Agreements, it would therefore seem that the New Mexico consolidated interim storage site could be dispensed with now and any plans abandoned. There is no such similar provision for the repository process in Title II.

How We Arrived at This Impasse

After more than 50 years of effort, the federal nuclear waste program in this country has failed to deliver a final resting place for highly toxic, radioactive waste that will be dangerous for millennia. Over the years, there have been numerous efforts to attribute the failure of the repository program to certain Senators, to Nevada Governors of both parties, to NRC Commissioners, and even to the public for failure to accept its part in disposing of nuclear waste. All of this is wrong.

Failure cannot be laid at the feet of any one person or entity or the public; rather, this defeat has many causes. Several agencies (including the EPA, the DOE, the NRC, and the U.S. Department of Justice (DOJ)) and Congress repeatedly distorted the process established in the NWPA, including for developing licensing criteria for a proposed repository. In each instance, such action weakened environmental standards rather than strengthening them, and always to ensure the site would be licensed, no matter the end result. These actions both precipitated and gave traction to ferocious resistance from Nevada, Tennessee, New Mexico, Washington, Texas, Louisiana, Mississippi, Utah, Georgia, Maine, Minnesota, New Hampshire, North Carolina, Virginia, Wisconsin, and Indian tribes. But even those actions are not the reason we remain locked in a virtual *cul de sac*, witness to repeated attempts to try and force the same result by the same fashion – *i.e.*, transferring the entirety of the nation's nuclear waste to an above ground parking lot in a resistant New Mexico, or to the technically inadequate attempt at a repository in Nevada.

Science & Politics Are Both Necessary

Nuclear waste remains a third rail of American politics for a singular reason – a deep misunderstanding of federalism and the necessary role of states in the process of solving this challenge. If you take one message from our appearance before you today, it is that there is another way to try and cut this Gordian Knot, but it must be done in a fashion that respects the extraordinary history of cooperative federalism in environmental laws.

We urge the Committee to appreciate the metamorphosis of Congressman Mo Udall's (D-AZ) NWPA, the organic subject of today's hearing. Indeed, NRDC views the original incarnation of the NWPA as a remarkable, nearly visionary piece of legislation that contained one tragic, fatal flaw: a deep misunderstanding of federalism and the necessary role of states. And that flaw is the single clear conclusion that we have drawn from the history of failures associated with nuclear waste.

As the Committee is aware, the enacted 1982 NWPA set forth obligations and duties for EPA, DOE and NRC, with Congressional oversight and checkpoints along the way. The law attempted to place science in the forefront and balance political power in a way that might allow for this fraught, difficult process of finding and developing disposal sites for nuclear waste. But, importantly, the NWPA never challenged or altered in any way the AEA's provision for exclusive federal jurisdiction over radioactive waste. Despite this baked-in oversight, the NWPA's attempt at the legal balancing act was unprecedented at the time and that observation remains true today. And as we all know, the balancing act was upset as the NWPA was repeatedly altered and the process was finally abandoned by the previous administration in 2009.

But why the repeated derailments? Some of my fellow witnesses here today suggest that "not in my backyard" (NIMBY) sensibilities and associated politics are responsible for the failure to license and open Yucca Mountain. But as noted at the outset – this is wrong. The deep misunderstanding of federalism and the necessary role of states at the heart of the NWPA just kept getting lost over the years. The federal exclusivity over nuclear waste regulation was simply presumed *a priori*, without consideration as to whether that might be at the root of the problem.

So how is the misunderstanding of federalism at the root of the problem? The relationship of the federal government to the governments of the 50 states that comprise our republic is the fundamental fact of American politics. Our political system has never easily digested or durably solved profound national problems like voting rights, health care, gun control, carbon restrictions, or the disposal of nuclear waste by either federal fiat or, conversely, by turning matters over to the states entirely.[11] And in every instance of national decision making on these and other complex issues, heavily compromised laws or regulations have taken into account the needs and perspectives of states.

Bedrock environmental laws reflect this fact. With the notable exceptions of the AEA (the organic act for nuclear power) and its progeny, the NWPA, there is federalist intention at the heart of environmental statutes and a role expressly reserved for the states. As examples, the Clean Water Act, Clean Air Act, and Resource Conservation & Recovery Act (RCRA) allow states authority to implement those air, water, and waste programs, respectively, in lieu of a federal program. States that obtain "delegated" authority from the

[11] For perspective on the ever-present interplay of the constitutional principles of federalism and equal sovereignty of the states and the extraordinary controversies that still attend such matters, *see* the 2013 landmark (5 votes to 4 votes) Voting Rights decision and its vigorous dissent, *Shelby County, Ala. v. Holder*, 133 S. Ct. 2612 (2013).

federal government must meet minimum federal standards (and the federal government retains independent oversight and enforcement authority). And generally, depending on state law, those delegated states can impose stricter requirements or different, but no less protective, regulatory mandates that meet the needs of the state in question. Nuclear waste should be no different, but under the AEA and the NWPA, it is different.

So, where do these observations leave us? It is NRDC's firm conclusion that Congress is right to take up these matters, that new nuclear waste legislation must be written, and that a new process must be created. Consistent with the expressed statements of so many in the Congress today, whatever results must be "consent based," concordant with President Obama's bipartisan BRC, and take into account the needs of the industry and their federal champions. But this time, any new legislation must also take into account the fundamental need for public and state acceptance and there is only one way to do that, as we explain next.

It Is Past Time to Normalize the Treatment of Nuclear Waste under Environmental Law

State consent and public acceptance of a nuclear waste solution will never be willingly granted unless and until power to make such a decision as to how, when and where such waste is disposed of is shared rather than decided by federal fiat. There is only one way that can happen consistent with the protective, cooperative federalism at the heart of environmental law.

Specifically, Congress must finally end the AEA's exemptions from environmental law. Our hazardous waste and clean water laws must have full authority over radioactivity and nuclear waste facilities so that EPA and – most importantly – the states can assert direct regulatory authority. This will necessarily alter the federalism oversight that has been central to the failure of the NWPA.

The NWPA's (and AEA's) misunderstanding of the importance of federalism is at the heart of the repository program's failure. If we don't find a way to give EPA and the states regulatory power over nuclear waste – and that is accomplished only by doing away with the environmental exemptions in the AEA – we will not solve this dilemma. Lack of consent from an unwilling host state selected in an expedient demonstration of legislative and administrative power over the (statutorily defined) powerless is a recipe for inaction and, ultimately, disaster in this country, whether the issue is nuclear waste or any other great public concern.

RDC's Specific Prescription & How to Get This Right

Five Recommendations to Get the Nuclear Waste Program Back on Track

We can dispose of nuclear waste and do so in a fashion that is both scientifically defensible and publicly accepted, but we cannot do so if we keep using the approach that has failed for over 50 years. To that end, NRDC urges Congress to – (1) recognize that geologic repositories must remain the focus of any legislative effort; (2) create a coherent legal framework before commencing any geologic repository or interim storage site development process; (3) arrive at a consent-based approach for nuclear waste storage and disposal via the fundamental change in law we described above; (4) address storage in a phased approach consistent with the careful architecture of former Senator Bingaman's S. 3469 (introduced in 2012); and (5) exclude delaying, proliferation-driving and polarizing closed fuel cycle and reprocessing options from this effort to implement the interim storage and ultimate disposal missions.

Rather than repeat mistakes of the last four decades, Congress must create a transparent, equitable process incorporating strong public health standards that are insulated from weakening those same standards when expedient to license a facility. Such a process can conclude with the licensing and operation of a suitable repository site (or sites) that can be effectively regulated under long effective environmental laws. We will briefly describe the criteria necessary for this path.

Recommendation 1 - Deep Geologic Repositories Are the Solution for Nuclear Waste and Must Remain the Focus

NRDC concurs with the long held, consensus recognition that our generation has an ethical obligation to future generations regarding nuclear waste disposal. Adherence to the principle of deep geologic disposal as the solution to this obligation is consistent with more than 60 years of scientific consensus. The decision to isolate nuclear waste from the biosphere implicates critical issues, including: financial security, environmental protection, and public health, and no other solutions are technically, economically, or morally viable over the long term. This is why NRDC strongly supports development of a science-based repository program that acknowledges the significant institutional challenges facing nuclear waste storage and disposal. Thus, in whatever legislation moves forward, we urge explicit adherence to the first purpose of the NWPA, 42 U.S.C. § 10131(b)(1), "to establish a schedule for the siting, construction, and operation of repositories that will provide a reasonable assurance that the public and the environment will be adequately protected from the hazards posed by high-level radioactive waste and such spent nuclear fuel as may be disposed of in a repository."

Recommendation 2 – Create a Coherent Legal Framework That Ensures the "Polluter Pays" before Commencing Any Repository or Interim Storage Site Development

To avoid repeating failures of past decades and consistent with the bipartisan BRC recommendations, both the standards for site screening and development criteria must be in final form before any sites are considered. Generic radiation and environmental protection standards must also be established prior to consideration of sites. To give this recommendation explicit and simple context, as well as a precise set of language to follow, former Senator Bingaman's 2012 legislative effort (S.3469, specifically in Sections 304, 305 and 306) set in place some of the necessary structures that could avoid repeating the failure of the Yucca Mountain process. Specifically, the bill would have directed EPA to adopt, by rule, broadly applicable standards for the protection of the general environment from offsite releases of radioactive material from geologic repositories. The bill also directed NRC to then amend its regulations governing the licensing of geological repositories to be consistent with any relevant standard adopted by EPA. Further, embedded in Senator Bingaman's bill was the requirement that the polluters pay the bill for the contamination created. This bipartisan concept has long history as bedrock American law and must remain in full force in any legislation.

These requirements and this phasing of agency actions in Senator Bingaman's bill were appropriate (*i.e.*, first EPA sets the standards and then NRC ensures its licensing process meets those standards) – and in the next recommendation we'll expand on how this coherent legal framework must be improved. But it is key that a coherent legal framework be in place before siting decisions get made. Unfortunately, recent iterations of nuclear waste legislation, including the items on offer today, ignore this wise sequencing, thus ignoring BRC's recommendation that new, applicable rules be in *final* form before site selection.

Congress should also direct that standards for site screening and development criteria be based on careful characterization of the radiation sources and resulting doses. The chief sources of radiation in high-level nuclear waste are the beta-decay of fission products like Cs 137 and Sr-90 and the alpha-decay of actinide elements like Uranium, Neptunium and Americium. Beta-decay is the primary source of radiation during the first 500 years of storage, as it originates from the shorter-lived fission products. Then alpha-decay becomes the dominant source after approximately 1,000 years. These radiation sources and doses must be considered to ensure a scientifically defensible legal framework for site selection.

Recommendation 3 – Develop a Consent-Based Approach for Nuclear Waste Disposal through a Fundamental Change in Law

The BRC Failed to Define Consent & Thereby Did Not Point the Way Forward

For all its laudable qualities, the 2012 BRC report did not accurately portray the fundamental problem facing how to finally solve our nuclear waste disposal challenges. The BRC should have explicitly stated – and we do so here today – that Congress, with its firm understanding of federalism, should legislate a role for EPA and the states in nuclear waste disposal by amending the AEA to remove its express exemptions of radioactive material from environmental laws.

State, local and tribal governments *must* be central in any prescription for a successful repository and waste storage program. Regrettably, current law has treated these relationships as dispensable afterthoughts, preempted from any meaningful power and authority over radioactive waste disposal sites. And H.R. 2699 and 3136 suffer the same malady.

Rather than address this problem head on, seven years ago the BRC chose to carefully skirt the matter in its report, while still noting that federal and state tensions are often central in nuclear waste disputes. We think this failure to squarely address the matter provides the continued impetus to ignore this elephant in the room. The BRC's Final Report states in pertinent part:

> We recognize that defining a meaningful and appropriate role for states, tribes, and local governments under current law is far from straightforward, given that the Atomic Energy Act of 1954 provides for exclusive federal jurisdiction over many radioactive waste management issues. Nevertheless, we believe it will be essential to affirm a role for states, tribes, and local governments that is at once positive, proactive, and substantively meaningful and thereby reduces rather than increases the potential for conflict, confusion, and delay.

BRC Final Report at 56 (citation omitted).

The first sentence above both makes an observation and states a fact. The observation is that defining a meaningful and appropriate role for states, tribes, and local governments under current law is far from straightforward. The fact is that the AEA provides for exclusive federal jurisdiction over many radioactive waste management issues. According to the BRC, the difficulty of defining a meaningful and appropriate role for states is a "given" because of the fact of exclusive federal jurisdiction.

So what did the BRC suggest Congress do about this? Do away with the explicit federal jurisdiction? Increase the exclusivity of the federal jurisdiction? Somehow argue that the problems can be addressed without altering the exclusive federal jurisdiction in some fashion? There is nothing so clear or direct in the text. Rather, the BRC's very next sentence is simply an aspiration, without any explicit recommendation addressing the "given" (*i.e.*, exclusive federal jurisdiction) that makes the process so difficult. The BRC simply noted that it is "essential to affirm a role for states, tribes, and local governments that is at once positive, proactive, and substantively meaningful." NRDC agrees with the aspiration, but plainly the BRC missed an important opportunity to address the fundamental roadblock to solving our nuclear waste problem.

Without fundamental changes in our current, non-consent based law that explicitly address what the BRC termed, "federal, state and tribal tensions," we will never approach closure and consent on transparent, phased, and adaptive decisions for nuclear waste siting. We now explore in more detail this decades-overdue change in the law.

NRDC's Prescription for Ensuring States' Authority – Remove the AEA's Exemptions from Environmental Law

As we stated at the outset (*supra* at 2), a meaningful and appropriate role for states in nuclear waste storage and disposal siting can be accomplished in a straightforward manner by amending the AEA to remove its express exemptions of radioactive material from environmental laws. The exemptions of radioactivity make it, in effect, a *privileged pollutant*. Exemptions from the Clean Water Act and RCRA are at the foundation of state and, we submit, even fellow federal agency distrust of both commercial and government-run nuclear complexes. Removing the exemptions would make the treatment of radioactive waste consistent with every other bedrock environmental law.

As the Subcommittee is aware, most federal environmental laws expressly exclude "source, special nuclear and byproduct material" from the scope of health, safety and environmental regulation by EPA or the states, leaving the field to DOE and NRC. In the absence of clear language in those statutes authorizing EPA (or states where appropriate) to regulate the environmental and public health impacts of radioactive waste, DOE retains broad authority over its vast amounts of radioactive waste, with EPA and state regulators then only able to push for stringent cleanups on the margins of the process. The NRC also retains far reaching safety and environmental regulatory authority over commercial nuclear facilities, with agreement states able to assume NRC authority, but only on the federal agency's terms.

States are welcome to consult with NRC and DOE, but the federal agencies can, and do, assert preemptive authority where they see fit. This has happened time and again at both commercial and DOE nuclear facilities. This outdated regulatory scheme is the focal point of the distrust that has poisoned federal and state relationships involved in managing and disposing of high-level radioactive waste and spent nuclear fuel, with resulting significant impacts on public health and the environment.

If EPA and the states had full legal authority and could treat radionuclides as they do other pollutants under environmental law, clear cleanup standards could be promulgated, and the Nation could be much farther along in remediating the toxic legacy of the Cold War nuclear weapons production complex. Further, we could likely avoid some of the ongoing legal and regulatory disputes over operations at commercial nuclear facilities. Indeed, the BRC Report discusses New Mexico's efforts to regulate aspects of the Waste Isolation Pilot Plant under RCRA as a critical positive element in the development of the currently active site (BRC Final Report at 21). Any regulatory change of this magnitude would have to be harmonized with appropriate NRC licensing jurisdiction over facilities and waste, and harmonized with EPA's existing jurisdiction with respect to radiation standards: but such a process is certainly within the capacity of the current federal agencies and engaged stakeholders. Some states would assume regulatory jurisdiction over radioactive material as delegated programs under the Clean Water Act or RCRA, and others might not. In any event, substantially improved clarity in the regulatory structure and a meaningful state oversight role would allow, for the first time in this country, consent-based and transparent decisions to take place on the matter of developing nuclear waste storage sites and geologic repositories.

Ending the anachronistic AEA exemptions does not guarantee a repository will be sited in the next few years. Indeed, expecting immediate progress on nuclear waste seems a fool's errand in light of the history. But ending these exemptions and providing RCRA authority for nuclear waste solves the most crucial matter for consent – the opportunity for meaningful, ongoing state oversight over nuclear waste. Any such statutory change bars the substantial likelihood of Congressional terms and modifications exacted from states (that might be willing to host a repository) years into a good faith negotiation on a site. Indeed, while it would be theoretically possible for a future Congress to revisit the AEA and re-insert exemptions from environmental law, it would have to do so in a manner that would remove jurisdictional authority from all states (or Congress would have to single out one state for special treatment). The difficulty of prevailing over the interest of all 50 states rather than simply amending legislation that affects the interests of just one state should be apparent. It is past time to normalize nuclear waste with the rest of environmental law and NRDC sees this as the key to developing a durable consent-based approach.

Recommendation 4 – Address Storage in a Phased Approach Consistent with the Careful Architecture of 2012's S. 3469

Efforts to initiate a temporary away-from-reactor storage facility – that are now, unfortunately, in process in H.R. 2699 and 3136 – must be inextricably linked with development of a permanent solution. This linkage, which is a crucial guard against a "temporary" storage facility becoming a permanent one, or essentially dictating the choice of a nearby site, should guide the legislative process. Consistent with the BRC's findings, a case can only be made for interim storage if it is an integral part of the repository program and not as an alternative to, or *de facto* substitute for, permanent disposal.

Specifically, the only way in which NRDC could see merit in a pilot project is in a hardened building,[12] located at one of the currently operating commercial reactor sites. These potential volunteer sites – operating commercial reactors – already have demonstrated "consent" by hosting spent nuclear fuel for years or decades. Far less of the massive funding that would be necessary in the way of new infrastructure would be required, and the capacity for fuel management and transportation is already in place, along with the consent necessary for hosting nuclear facilities in the first instance. Further, Congress would avoid entirely the ferocious fight that is sure to ensue with New Mexico and Texas citizens (and as happened with Utah and Tennessee) if they continue down the road with the DOE and the existing license applications in those states.

---

[12] An example of such a hardened building is the Ahaus facility in Germany.

Rather than prematurely bypassing a careful, consent-based process that can arrive at protective, publicly accepted and scientifically defensible solutions, we have urged NRC and industry to focus spent fuel storage efforts on ensuring that all near-term forms of storage meet high standards of safety and security for the decades-long time periods that interim storage sites will be in use. Congress could legislatively direct such efforts and would be wise to do so.

Recommendation 5 – Exclude Unsafe, Uneconomic Closed Fuel Cycle and Reprocessing Options from This Effort

Both the BRC Recommendations and all the subsequent legislative iterations (including those under discussion today) have, for the most part, wisely resisted inclusion of support for reprocessing, fast reactors, or other closed fuel cycle options as a corollary to new nuclear waste policy. We agree with relevant BRC findings, that there are "no currently available or reasonably foreseeable" alternatives to deep geologic disposal.[13] As Senator Bingaman noted in 2012 at the outset of legislative efforts subsequent to the BRC process, "even if we were to reprocess spent fuel, with all of the costs and environmental issues it involves, we would still need to dispose of the radioactive waste streams that reprocessing itself produces and we would need to do so in a deep geologic repository."[14] At no point should this evolving nuclear waste process include support for closed fuel cycle options.

Conclusion

On one thing I hope we can all agree; the history of the federal nuclear waste program has been dismal. But decades from now others will face the precise predicament we find ourselves in today if Congress again tries to push through unworkable solutions contentiously opposed by states, lacking a sound legal and scientific foundation, and devoid of wide public acceptance and consent. Efforts to quickly restart the abandoned Yucca Mountain licensing process or fast track an interim storage facility will not work, lead to years of litigation, and thereby derail needed efforts to find scientifically defensible disposal sites. Unless and until Congress fundamentally revamps how nuclear waste is

---

[13] BRC Final Report at 100.

[14] *See*, *Previewing the Nuclear Waste Bill*, Remarks by Chairman Bingaman to the Bipartisan Policy Center, June 6, 2012, online at https://www.energy.senate.gov/public/index.cfm/democratic-news?ID=490349a4-4b5e-4ac2-83e7-6e9a54c7aaf0.

regulated and allows for meaningful state oversight by amending the AEA to remove its express exemptions of radioactive material from environmental laws, we're doomed to repeat this dismal cycle until a future Congress gets it right.

We deeply appreciate the opportunity to testify today and I am happy to answer any questions.

NRDC Attachment 1

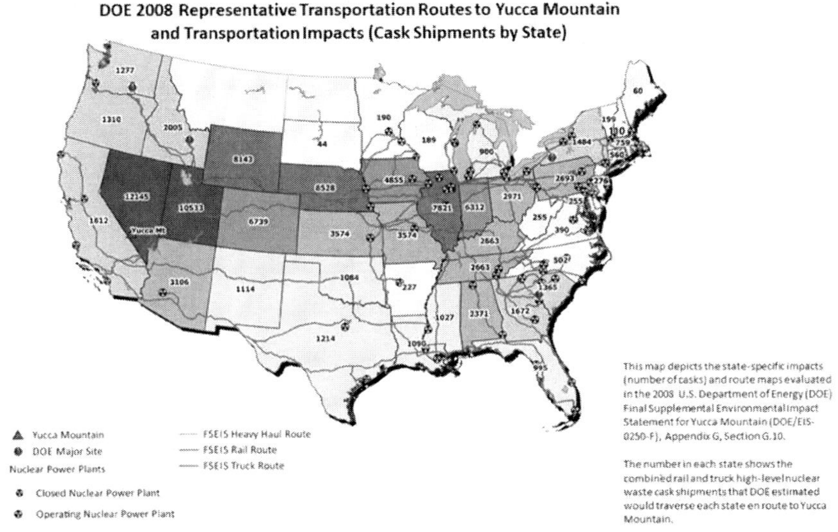

NRDC Attachment 2

State of New Mexico

Michelle Lujan Grisham
Governor

June 7, 2019

The Honorable Rick Perry
Secretary
U.S. Department of Energy
1000 Independence Avenue, SW
Washington, DC 20585

The Honorable Kristine Svinicki
Chairman
U.S. Nuclear Regulatory Commission
Mail Stop O-16B33
Washington, DC 20555-0001

Dear Secretary Perry and Chainnan Svinicki:

I write to express my opposition to the proposed interim storage of high-level radioactive waste in the state of New Mexico. The inter im storage of high-level radioactive waste poses significant and unacceptable risks to New Mexicans, our environment and our economy. Furthermore, the absence of a permanent high-level radioactive waste repository creates even higher levels ofrisk and uncertainty around any proposed inter im storage site.

As you know, the Nuclear Regulatory Commission (NRC) is evaluating the issuance of a 40-year license to Holtec International for a consolidated interim storage facility in southeastern New Mexico. As proposed, this facility would store spent nuclear fuel (SNF) and reactor-related materials greater than low- level radioactive waste.

A facility of this nature poses an unacceptable risk to New Mexicans, who look to southeastern New Mexico as a driver of economic growth in our state. New Mexico's agricultural industry contributes approximately $3 billion per year to the state's economy, $300 million of which is generated in Lea and Eddy Counties, where the proposed faci lity is to be sited.

Further, the Pennian Basin, situated in west Texas and southeastern New Mexico, is the largest inland oil and gas reservoir and the most prolific oil and gas producing region in the world. New Mexico's oil and natural gas industry contributed approxima tely $2 billion to the state last year. According to the U.S. Energy Infonnation Administration (EIA), Lea County and Eddy County were ranked the second and sixth oil-producing counties in the country, respectively, earlier this year, with production continuing to mcrease.

Establishing an interim storage facility in this region would be economic malpractice. Any disruption of agricultural or oil and gas activities as a result of a perceived or actual incident would be catastrophic to New Mexico, and any steps toward siting such a project could cause a decrease in investment in two of our state's biggest industries. For those reasons, the New Mexico Cattle Growers' Association, the New Mexico Farm and Livestock Bureau and the Pennian Basin Petroleum Association have all sent me letters opposing high-level waste storage in southeastern New Mexico. I have attached their letters for your review.

In addition to significant economic concerns about th.is project's potential impact on agriculture and the oil and gas industry, I am concerned about the financia 1 burden it could place on the state and local communities. Transporting material of this nature safely requires both well-maintained infrastructure and highly specialized emergency response equipment and persom1el that can respond to an incident at the facility or on transit routes. The state of New Mexico cannot be expected to support these activities.

Finally, given that there is currently no permanent repository for high-level waste in the United States, any interim storage facility will be an indefinite storage facility. Over this time, it is likely that the casks storing SNF and high-level wastes will lose integrity and will require repackaging. Any repackaging of SNF and high-level wastes increases the risk of accidents and radiological health risks. Again, New Mexicans should not have to tolerate this risk.

Given the potential for adverse impacts to public health, the environment and our economy, I cannot support the interim storage of SN.F or high-level waste in New Mexico.

I thank you for your consideration of these concerns and look forward to your reply.

Sincerely,

Michelle Lujan Grisham
Governor

* * *

Mr. TONKO. Thank you, Mr. Fettus.

We now recognize Mr. Barrett for five minutes, please.

Statement of Lake Barrett

Mr. BARRETT. Thank you, Chairman Tonko, Ranking Member Shimkus, and distinguished members of the committee. It is an honor to appear before you today on the importance and urgency of moving forward to remove spent nuclear fuel from shutdown reactor sites as soon as possible while providing our grandchildren with a permanent disposal solution.

I speak to you today from the perspective of a former Department of Energy civil service executive who spent nearly two decades trying to implement the laws of the United States in this area. I retired from DOE in the 2002 after completing the statutory Yucca Mountain site recommendation process. This process included Nevada's disapproval of the Yucca Mountain site and subsequent bipartisan congressional override votes of 307 here in the House and 60 votes in the Senate.

The need for Federal performance to remove spent fuel from reactor sites is growing faster as more power reactors shut down, leaving fuel stranded at dozens of sites across the country. Meanwhile, the lawful DOE program created to solve this problem is stuck in limbo accomplishing nothing now because Nevada delegation has skillfully politically stopped program funding.

To address these issues, the country needs to keep its primary focus on a realistic geologic repository program and add integrated consolidated interim storage as an important supplement for a more timely removal of spent fuel from shutdown reactors.

In my view, H.R. 2699 moves in that direction as it contains the necessary elements of both the repository and interim storage facility. Specifically, I urge congressional action to complete the almost finished Yucca Mountain licensing proceeding. Yucca Mountain is the most studied piece of ground on the planet with separate teams of DOE and Nuclear Regulatory Commission staff scientists concluding that all million-year safety requirements are met. Although Nevada disagrees, it is time, before impartial judges, for a safety decision.

Regardless of Yucca Mountain progress or not, the Federal Government should engage Nevada to discuss various empowerment and partnership options that further respect Nevada's host State status. It is time we break the repeating win-lose, lose-win cycle that has frustrated both Nevada and the Nation for decades. Both sides have much to gain and both sides have much to lose.

I personally believe a win-win scenario can be mutually developed to conserve the needs of both. I am confident that the Yucca Mountain repository will succeed on its scientific and regulatory merits. It is possible, however, that it may fail or be endlessly delayed by political obstructionism. And even if Yucca Mountain does go forward under current law, another repository will be needed. Thus, an alternative repository program should be promptly started.

An integrated consolidated interim storage facility should be added to the Nation's spent fuel management program because it would provide a bridge from the many dozens of temporary fuel storage locations across the country to an eventual ultimate geologic repository site in a more timely manner.

However, there are significant siting challenges for an interim storage facility based on experience because, despite local community support, host State level leadership has generally stopped these efforts. A major reason for the blockage was that any interim storage facility without a companion realistic geologic disposal program would just become another indefinite storage facility. It is for this reason that I believe the existence of a meaningful repository program will be a necessity to enable the development of a consensus consolidated interim storage facility anywhere in this country.

Speaking as a grandparent and as an engineer, it is simply irresponsible to saddle our children, grandchildren, and future generations with spent fuel sitting in thousands of canisters in dozens of stranded storage sites scattered across this country on our rivers, lakes, and seashores, with seemingly endless financial liabilities and no place to go. It is time we act to remove spent fuel from the coast of Maine to the coast of California and from our Great Lakes and river systems in between. It is time to step up and take responsibility for decisions we made six decades ago to produce nuclear fuel and three decades ago to develop a geologic repository for an ultimate safe disposition of that fuel.

Thank you very much.

[The prepared statement of Mr. Barrett follows:]

Submitted Testimony from, Lake H. Barrett, House Commerce and Energy Committee, Environment and Climate Change Subcommittee Hearing, Cleaning up Communities: Ensuring Safe Storage and Disposal of Spent Nuclear Fuel, June 13, 2019, 2322 Rayburn House Office Building

Chairman Tonko, Ranking Member Shimkus and distinguished members. It is an honor to testify before you today on the importance and urgency of moving our country forward with a safe environmentally protective nuclear waste management program that supports the timely removal of spent nuclear fuel from commercial reactors, with a priority towards shutdown reactor sites, while also providing our grandchildren with a permanent disposal solution.

I speak to you today from the perspective of a former Department of Energy (DOE) civil service executive who spent nearly two decades trying to implement the laws of the United States in this area. I directly reported to five different Secretaries of Energy under three different administrations, serving as Principal Deputy Director of the Office of Civilian Radioactive Waste Management for nearly a decade and as the Acting Director longer than any Senate-confirmed Director. I retired from federal service upon completion of most of the statutory Yucca Mountain repository site scientific work, the Presidential Site Recommendation, and the Congressional override of the State of Nevada disapproval of the site in 2002.

I urge the Committee adopt legislation that supports:

- Completing the almost finished Yucca Mountain NRC licensing proceeding
- Encouraging DOE and Nevada to engage in good faith discussions to resolve Nevada's concerns with empowerment and partnership options that can serve the needs of both
- Starting an additional geologic repository development program
- Authorizing Integrated Consolidated Interim Storage
- Moving forward promptly

The framework for our nation's nuclear waste management program for commercial spent nuclear fuel and defense high-level radioactive wastes is detailed in the Nuclear Waste Policy Act of 1982, as amended in 1987 (NWPA). Many of the exact issues that the Congress is confronting today, such as balancing host state interests with national needs and shutdown reactors, were debated then and processes were established in law to address each issue. However, the implementation of that framework has been delayed for many reasons and is currently stymied by the lack of Congressional appropriations.

Over the decades, some of the delays have been caused by additional extraordinary scientific safety and environmental protection studies and investigations to demonstrate compliance with the conservative Environmental Protection Agency (EPA) and Nuclear Regulatory Commission (NRC) million-year requirements, extra public participation steps, and legal challenges.

However, the most significant source of delay has been, and continues to be, an effective Nevada based politically driven effort to prevent Congressional funding coupled with an unwillingness to proceed.

The Nevadan Congressional members argue obstruction is appropriate because: they believe that the 1987 NWPA amendment was patently unfair by requiring that only the geologic repository site in their state was to be evaluated, claim that the site is not scientifically sufficient, and that out-of-state wastes should not be forced into their state. They also have argued that they do not have enough control over a Federal repository facility, and that they should have an absolute veto authority over a federal nuclear waste geologic repository in their state.

In addition, some interest groups want to effectively abandon the 1982 statutory framework by restarting a new program with an undefined multi-decade process to find a different unspecified geologic repository site that requires consent from state and local governments, and affected tribes. This is in addition to the already challenging necessary rigorous scientific safety and environmental protection work required. Others wish to focus mostly on centralized interim storage facilities and reduce the focus on developing a geologic repository facility.

I believe that the DOE should further engage with the Nevadan's to address their concerns. However, I do not believe that the near abandonment of the original NWPA balancing framework is in the best interest of our nation to move forward. Adjustments to existing law, such as many contained within the subject bills of this hearing, are appropriate to better enable the government to meet federal statutory and contractual responsibilities, and indeed our moral responsibilities for nuclear waste management. The most critical need, however, is to finish the last step of the NRC Yucca Mountain licensing process so that the nation, Nevada included, can have a fair and impartial assessment of the safety of the facility and then- if all safety requirements are satisfied-decide whether to build it, or not.

The federal waste management program is now stuck in limbo and going nowhere, while the need for action to remove spent nuclear fuel from reactor sites is growing quickly as power reactors are nearing their end of life and more are being forced to close early due to economic conditions. Although all reactor sites can store wastes safely for many decades, reactor sites were selected for power production-located relatively near population centers, on our coasts, lakes, and rivers, and not selected for long term waste storage or becoming *de facto* permanent nuclear waste storage sites. Today there are two dozen shutdown or announced shutdown sites and the list is growing. The communities near these reactors never agreed to be long-term waste storage sites and for shutdown reactors, the presence of these wastes precludes site reuse for commercial or public purposes and is a significant burden to everyone.

The nation's electricity consumers have already paid the Federal government for the disposal of their reactor spent fuel in their electricity bills over the past 30 years. The value of this federally held fund is approximately $41 Billion dollars currently. And in addition, federal taxpayers also are legally required to pay damage costs of over two million dollars each day for extra unnecessary reactor storage costs. These taxpayer funded damage costs are continually rising and will be in the total of many tens of Billions of dollars range in the relatively near future if federal performance does not happen soon.

Not having a meaningful national spent fuel disposal program for our current reactor fleet can also negatively impact the public confidence in our national plans for the development and deployment of new clean air advanced nuclear energy systems, e.g., small modular reactors. These new reactors are designed to minimize wastes, so the addition will be very minor relative to the amounts already in existence. However, the current lack of progress on waste disposal does hinder our nation's ability to achieve these major environmental climate protection advances by allowing critics to claim that there is no solution for new future nuclear wastes and that these clean safe energy sources should not be supported.

Given this current situation, it is important that the Congress provide funding and additional guidance to the DOE, or any successor organization that Congress may decide by law, to move forward with an effective integrated spent nuclear fuel management program. It is my view that H.R.-2699 moves in that direction as it contains the necessary elements of both permanent repository and interim storage programs.

To address the current national challenges, the country needs to keep its primary focus on a realistic geologic repository program and add consolidated interim storage as an important supplement to provide greater assurance for expedited spent fuel removal from priority shutdown reactors. Having a realistic path forward for the ultimate disposition of spent fuel in a geologic repository, such as at Yucca Mountain, is likely going to be a necessity for sustaining the realization of a consolidated interim storage facility somewhere in this country.

Yucca Mountain Repository

I strongly support finishing the Yucca Mountain NRC statutorily directed adjudicatory licensing process to address the scientific concerns of the State of Nevada and others that have standing in the hearing. Nevada, and others, have a right to have impartial independent judges hear the facts and allow the regulatory process to finish. This is not a decision to build the repository since that requires many additional approvals and steps. Even if a Yucca Mountain license is not ultimately granted, the nation will have learned many valuable regulatory lessons from a even failed Yucca Mountain experience that can be applied to a different site in the future.

Yucca Mountain is the most studied piece of land in the world. The science behind Yucca Mountain was conducted to the highest standards, has withstood countless national and international peer reviews, and was also overseen by a Presidentially appointed Nuclear Waste Technical Review Board which is composed of National Academies of Sciences nominated scientists. In addition to the many billions of dollars of DOE safety work, the NRC staff has spent over thirty years and approximately $600 million dollars doing detailed independent evaluations of all the safety issues. The NRC staff has issued a formal Safety Evaluation Report concluding that the site meets all regulatory safety and environmental protection requirements. While all technical issues have been addressed, this significant national investment (~$15B) will not be complete until the final stage of the public adjudicatory hearing process of the Nevada safety concerns is finished. This effort should not be lost by default.

In the early 1980s, Congress actively debated the proper balance of states' rights and federal need during the development and enactment of the NWPA. Although seldom mentioned these days, the NWPA provided for the development of a negotiated agreement between DOE and the host state-Nevada chose not to engage. Nevertheless, in accordance with law, DOE provided funding to Nevada to be involved in the scientific and regulatory processes for Yucca Mountain. The DOE provided the State of Nevada and their university system over $200 million dollars during the Yucca Mountain evaluation and licensing process. Most importantly, Nevada had the right to "disapprove" the Presidential Site Recommendation and that the site would have to be abandoned unless there was an override by both the houses of Congress. Nevada did disapprove the site; however, the disapproval was overridden by the House (306-117) in May 2002 and the Senate (with 60 votes to proceed to a voice override vote) in July 2002. However, although the original statute did provide measures for State participation and disapproval power, Nevada does not consider them sufficient and has been quite successful in continuing to stall progress through political means.

It is important to note that the local host government of the Yucca Mountain site, Nye County, as well as eight other surrounding local county governments (a majority of Nevada counties) support the continuation of the Yucca Mountain licensing and construction of the facility, if it is demonstrated to be safe under NRC rules.

To address Nevada's concerns, the federal government should engage Nevada to discuss various empowerment and partnership options that further respect Nevada's host role needs while also respecting national needs. It is time that we break to the repeating "win-lose; lose-win" cycle that has frustrated both Nevada and the nation for decades. Both sides have much to gain and much to lose. I personally believe a "win-win" scenario can be mutually developed that can serve the needs of both, but only if both sides are willing to engage in a constructive dialog.

Additional Repository Program

While the Yucca Mountain repository should succeed on its scientific and regulatory merits, it is entirely possible that it may fail (or be endlessly delayed) due to political obstructionism, and the nation should prepare for that potential outcome. Geologic disposal is the ultimate need, as supported by long standing national and international consensus reports. Additionally, the federal government has committed to siting such a permanent disposal facility in return for electric customer payments into the federal Nuclear Waste Fund. Therefore, any "alternatives" to the proposed Yucca Mountain repository must focus on permanent disposal in a deep geologic repository.

The NWPA has a requirement for both a first and second repository. So even if Yucca Mountain were to go forward, there is a current non-technical statutory capacity limitation that requires at least a second repository. Therefore, a search for another geologic repository somewhere within the United States should start sooner rather than later because it will be a long (many decades), difficult, and expensive process.

The Blue-Ribbon Commission on America's Nuclear Future, and others, have recommended a shift to a consent-based approach for siting a repository. The consent-based siting approach has made progress in some smaller countries that basically have only national and local government structures.

Consent-based siting has not been as successful in larger countries with regional (state) governments like Japan, Germany and the United Kingdom, which have had serious difficulties at the state/regional level, similar to what we have encountered here. If the United States had only national and local government involvement, this problem likely would have been solved years ago.

A consent-based siting approach would be ideal, and I wish it could succeed, but based on past experiences I personally do not believe it can succeed in our complex state/federal government system now or in the future. Nonetheless, a repository siting process, consent-based or otherwise, should be started soon to address the possibility of not having Yucca Mountain as a nuclear waste repository or to address the current statutory limitations if Yucca Mountain proceeds.

Initiating a new consent-based repository siting process will face substantial challenges. Consent must be defined, e.g., who would be involved (governor, legislature, neighboring jurisdictions, transportation corridor jurisdictions, etc.) and what public processes would be used, e.g., referendums? New generic EPA standards and NRC regulations will have to be developed since current standards are either Yucca Mountain specific or not appropriate for future use. These things can all be done, but it will not be an easy, fast, or inexpensive endeavor.

The siting of another repository will of course require having a strong, complex scientific safety and environmental protection technical program as well as an integrated program to create social/political host approval. The DOE tried to do this with the NWPA second repository program back in the mid-1980s, but that program was terminated due to opposition in the states that were under consideration. Although DOE tried to address state level concerns at that time, the program was heavily scientifically based in earth sciences and was incapable of dealing with the emotional and eventual political resistance. Nonetheless, that program did identify potential geologic formation regions for a repository in 36 states across the nation. Eventually a technically suitable site will be needed in some state.

Integrated Consolidated Interim Storage

I believe the addition of an integrated consolidated interim storage facility into the nation's spent fuel management program is warranted and very important. However, there is much to learn from the past, as this nation has tried and failed to create consent based interim storage facilities many times in the past decades. Despite local community host support, host state leadership has routinely stopped the efforts. In the mid-1980s the state of Tennessee stopped the DOE proposal for a storage facility near Oak Ridge. In the early 1990s two federally appointed Nuclear Waste Negotiators (a Republican followed by a Democrat) tried for six years to negotiate a consensus repository or interim storage site location. However, local efforts to consider hosting a storage facility were blocked by state-level concerns. For example, the Governor of Wyoming stopped a Fremont County effort and the New Mexico delegation stopped a Mescalero Apache effort. In the early 2000s the State of Utah stopped the NRC-licensed Private Fuel Storage facility on the reservation of the Goshute Indian tribe.

Currently, with strong local support, a proposed private interim storage site in south east New Mexico is undergoing NRC licensing review. However, just last week, New Mexico Governor Grisham expressed her written opposition to that facility within her state.

In all of these cases, from the 1980s to the present, a major concern of the host state was that a national geologic repository would not come into existence and that any interim storage facility would become a permanent indefinite duration waste storage site. It is for this reason, that I believe the existence of a meaningful realistic repository program will be a necessity to enable the development of a consensus consolidated interim storage facility anywhere in this country.

Developing an integrated consolidated interim storage facilities will be a siting challenge, but I firmly believe it is doable and desirable in order to provide a bridge from the current many dozens of temporary nuclear waste storage locations across the country to an eventual ultimate geologic repository site by providing a diversity and redundancy of options for a more timely fuel receipt. Although spent nuclear fuel is being safely stored at reactor sites today, these sites were never meant to long-term waste storage sites and none of the site communities or states ever consented to indefinite storage. This is unnecessarily costing our nation tens of billions of dollars.

Speaking as a grandparent, as well as an engineer, it is simply irresponsible to saddle our children, grandchildren and future generations with spent nuclear fuel sitting in thousands of canisters in dozens of temporary storage locations scattered across the country with seemingly endless financial liabilities with no place to go It is time to act to remove spent fuel from the coasts of Maine to the coasts of California and from our Great Lakes and river systems in between. It is time to step up and take responsibility for decisions we made six decades ago to produce nuclear fuel and three decades ago to develop a geologic repository for the ultimate disposition of that spent nuclear fuel.

Thank you for your attention.

* * *

Mr. TONKO. Thank you, Mr. Barrett.

We have concluded witness opening statements. We will now move to member questions. Each Member will have five minutes to ask questions of our witnesses.

Looking at the number of colleagues here, I am reminded that we are to have votes around 11:30. And those votes are going to run for hours. So I would ask that our witnesses offer succinct answers to these questions, and I encourage my colleagues to offer, again, succinct questions. So we will be really tough with the gavel today because, in fairness, I want everyone to get their questions in. We will have, then, about 45 to 50 minutes, and we will not return. So thank you.

We will now move to questions. I will begin with that process.

As I mentioned, this is a complex and difficult challenge, but I also believe that we have a responsibility to acknowledge that this waste exists and needs to be dealt with.

I would ask everyone on the panel, what is the one thing that can—that Congress can do that will result in the most progress in addressing our waste challenge?

Ms. Korsnick, why don't we begin with you?

Ms. KORSNICK. Well, I think the best thing we can do is to move forward with the licensing process with Yucca. I think that is the linchpin. It started. It is the law. The process is underway. We need to complete it.

Mr. TONKO. Thank you. Mr. Halstead.

Mr. HALSTEAD. The most important thing is to extend consent to the State of Nevada regarding Yucca Mountain, as is proposed in the legislation that has been put in by Nevada's delegation.

And I would tell you the opposite. The worst way to making progress is H.R. 2699.

Mr. TONKO. Thank you. Mr. Keyser.

Mr. KEYSER. I would echo Ms. Korsnick and say to go ahead and have the hearings—I would echo Ms. Korsnick and say that speaking up and considering the licensing of the DOE application for Yucca Mountain.

Mr. TONKO. Thank you. Thank you. Mr. Fettus.

Mr. FETTUS. To get progress started, remove the environmental exemptions in the Atomic Energy Act.

Mr. TONKO. Thank you. Mr. Barrett.

Mr. BARRETT. Complete Yucca Mountain licensing and use strong carrots and sticks to force the State of Nevada and the Federal Government to negotiate a win-win solution.

Mr. TONKO. And, Ms. Korsnick, as more reactors go through decommissioning, how will that exercise increase the urgency to resolve this standoff?

Ms. KORSNICK. Well, obviously, we have more waste that needs to be disposed of. So it only challenges it further.

When we talk about nuclear of the future, and there are wonderful opportunities for nuclear of the future, one of the first questions that we get is what about the waste. So the inability to have this solution really is an Albatross around the neck of the nuclear industry.

Mr. TONKO. OK. I will ask everyone on the panel, and perhaps this time begin with Mr. Barrett, what role should consolidated interim storage play?

Mr. BARRETT. I believe it should supply a supporting role to the ultimate disposition of a geologic repository, a very important supportive role.

Mr. TONKO. Thank you. Mr. Fettus.

Mr. FETTUS. We don't object to the concept. We do object to trying to push it forward without a reset that gives the cooperative federalism of our environmental laws control so States can actually consent or not consent.

Mr. TONKO. Thank you. Mr. Keyser.

Mr. KEYSER. Yes. I mean, we believe that there ought to be consent in the process, but it needs to happen. But we need a national repository as a backdrop. You can't have temporary facilities without them knowing they are going to be temporary facilities.

Mr. TONKO. Right.

And, Mr. Halstead, please.

Mr. HALSTEAD. It is necessary, because there is not going to be a repository for at least 20 and possibly 25 years, whether it is Yucca Mountain or something else. But it is important that there be an integrated approach so that the fuel that is packaged at the reactors can be accepted at the interim storage facilities and then eventually at a repository.

Mr. TONKO. Thank you.

And, finally, Ms. Korsnick, please.

Ms. KORSNICK. Yes. We agree that interim storage is an important step in the overall process to getting to long-term repository.

Mr. TONKO. And would interim storage, if opened ahead of a permanent repository, help limit taxpayers' liability through the judgment fund? Anyone? Yes, Mr. Barrett.

Mr. BARRETT. Yes, it could, if it is part of the integrated Federal system.

Mr. TONKO. OK. Anyone else want——

Mr. FETTUS. I don't think it will because I don't think you are going to have an integrated Federal system. I think you are going to have a de facto parking lot.

Mr. TONKO. Thank you.

Ms. KORSNICK. It doesn't meet the Federal obligation of receiving fuel, so I don't believe that it has a significant impact on the judgment fund.

Mr. TONKO. Thank you. And, Mr. Halstead.

Mr. HALSTEAD. The single major problem facing the program is not Yucca Mountain, not DOE, but the unresolved issue of how to manage the Nuclear Waste Fund, whether to reinstate the fee, and then how to deal with the way the Congress handles this in annual appropriations, and eventually what happens with the corpus.

So one thing I think that Mr. Shimkus and I agree on, I don't agree with his answer, but it is certainly the right question.

Mr. TONKO. Thank you very much.

With that, I will recognize Mr. Shimkus, the subcommittee ranking member, for five minutes to ask questions.

Mr. Shimkus.

Mr. SHIMKUS. Thank you, Mr. Chairman.

Ms. Korsnick, you note ratepayers contribute what amounts to a $41 billion balance in the Nuclear Waste Fund. And so, when Congress chooses not to spend money on the nuclear waste program, it is basically keeping ratepayer money for other purposes.

Would that be fair?

Ms. KORSNICK. No.

Mr. SHIMKUS. While Congress is not spending ratepayer money on the program they paid for, it is spending taxpayers money for its failure to move forward. Is it correct that these taxpayers funded liability payments are not appropriated by Congress but automatically paid out of the judgment fund?

Ms. KORSNICK. That is correct.

Mr. SHIMKUS. And the Congress doesn't really see that their constituents had been paying $800 million a year for failure to move forward, correct?

Ms. KORSNICK. That is correct.

Mr. SHIMKUS. Would you agree it is only going to get worse with each day of delay in appropriating funds to complete the licensing process?

Ms. KORSNICK. That is correct. It only gets worse.

Mr. SHIMKUS. That is why we need funding reform, and that is part of this debate.

Mr. Barrett, under the Nuclear Waste Policy Act, we gave Nevada the tools, which really means the money, they need to oppose the Yucca licensing process at the NRC. And I think that is instructive. And part of this whole process, the money that Nevada is using to dispute the science that I had directed in the opening statement came from ratepayers.

Mr. BARRETT. Yes, sir, it did. Nevada received about $200 million to the State of Nevada in the university system to follow the Yucca Mountain project, which they have done.

Mr. SHIMKUS. We also paid for all the scientific review and analysis that I showed, and for those who came late, those boxes over there, and that review, the white binders, are the NRC analysis. We have also put money forward to do that. Is that correct, Mr. Barrett?

Mr. BARRETT. Yes, sir.

Mr. SHIMKUS. What do we say to ratepayers if we just walk away from the Yucca Mountain licensing review and allow unadjudicated contentions to stand by default?

Mr. BARRETT. It is embarrassing.

Mr. SHIMKUS. It is embarrassing. Is that fair to the ratepayers?

Mr. BARRETT. In my view, no.

Mr. SHIMKUS. So, I want to—for the folks here and for my colleagues, this is where we are at. We are acting under a law that is the current law of the land. We did the science, $10 billion over 30 years. We gave Nevada the money to contest that. The last part of this process is to go before judges who are scientists to address the contentions that Nevada has that says that those boxes of science and that review by the NRC is not adequate enough.

For us to move forward and for—and, really, NRC too who prides themselves on science, we ought to an adjudicate the science, don't you agree, Ms. Korsnick?

Ms. KORSNICK. Absolutely. For Nevada to get their day in court, that is the part of the process that we are in right now. Let's take it through the licensing process and let the judges hear it.

Mr. SHIMKUS. Mr. Keyser.

Mr. KEYSER. Yes. Absolutely. We think it ought to move through the licensing process and have a fair review. And if that review turns up something we don't understand, we can reevaluate at that time. But a hearing is, I think, just in this case.

Mr. SHIMKUS. And Mr. Barrett.

Mr. BARRETT. Yes, sir.

Mr. SHIMKUS. So, Mr. Barrett, let me turn to you.

What if Nevada is right? What if they say—what if they go before these judges who are scientists, and when they lay out their contentions, and these judges who are scientists say it is not safe, per the current law of the land, the law, what happens?

Mr. BARRETT. If the judges say they would reject the license application and it would stop there and it is back here, we would have to basically say this site will not go forward, and we will have to find another alternative.

Mr. SHIMKUS. So, let me go through.

What do you mean by another alternative?

Mr. BARRETT. We would have to find another repository site somewhere of this country. I would note that there was a DOE second repository program in mid-1980s that was politically terminated. It identified suitable geologic regions in 36 States across this country. Something would have to go back with that, consent or not, and decide what to do, or possibly a different approach as Mr. Fettus talked about.

Mr. SHIMKUS. So, in the original—or our Founding Fathers of this whole debate, they threw out a wide net. And basically—to have a long-term geological repository, which is the scientific consensus for long-term nuclear waste, we have to start the process all over. And every State would have the opportunity to maybe welcome a nuclear waste long-term repository.

Mr. BARRETT. Yes, sir.

Mr. SHIMKUS. Thank you.

Mr. TONKO. The gentleman yields back.

I ask that our witnesses—our press corps has indicated they can barely hear some of the answers. So, if you could all just move in closer, please, to the mike.

The Chair now recognizes Representative Peters for five minutes.

Mr. PETERS. Thank you, Mr. Chairman.

I want to ask a question of Mr. Fettus. You addressed political things. You didn't actually discuss any of the technical issues.

Mr. Halstead posited that the defect in Yucca was that, even though we could talk about what you could engineer to hold nuclear waste, the geological setting was not adequate.

Does NRDC have an opinion on the whether the geologies at Yucca or in general is—

Mr. FETTUS. We do. We do. We think it is fraught with a series of significant technical questions. We think it is more leaky than was originally thought in the 1980s when it was chosen for, as Mr. Halstead said, political reasons. We also think there are significant questions whether it could ever survive the licensing process in any meaningful fashion.

Mr. PETERS. OK. Do you have an opinion—does NRDC have an opinion about places in the United States where, from a geological perspective, it would be appropriate to permanently dispose of nuclear waste?

Mr. FETTUS. You know, yes, in that we think that the process that Mr. Barrett just referenced, the U.S. Geological Survey process from the mid-1980s that did look at 36 States it came up with, and Mr. Barrett can correct me if I am wrong, dozens of locations that could potentially be suitable.

My point and why I spent the time on the institutional process— and I don't really see it as politics; I see it as the institutional framework for how to go forward—is you won't get to that scientific review, a meaningful deep review of what is a suitable site, until you reapportion the power consistent with our environmental laws.

Mr. PETERS. To me, that is obviously wrong. That is obviously wrong. I don't know if you can cite me any State that want to begin this.

Mr. FETTUS. Yes, I can. Who is asking for this? Which State?

MThis is why it won't work.

OK. At the end of the Obama administration—and I really appreciate the question—at the end of the Obama administration, they wanted to look at the viability of deep borehole disposal in South Dakota. South Dakota—red State South Dakota—they drew up the barriers so fast, so hard. And I would encourage you to go talk to DOE about what happened. It did not go well.

The reason why it did not go well was because any time the Energy Department Starts to look for it, a State thinks it could be entirely on the hook for everything. And until you change that, until consistent with environmental law, States have the ability to set the terms by which they could accept some portion, we will not get forward progress. You will have what happened with Nevada. We even have——

Mr. PETERS. But what you are saying is—we know Nevada doesn't want it. South Dakota didn't want it either. Tennessee didn't want it.

Mr. FETTUS. We don't know what will happen, honestly, if you actually reset the process where people are not on the hook for everything. And that is the key point.

Mr. PETERS. So what would be the terms with which you would approach South Dakota and say, "We are not going to give you everything; we are going to give you half of it"?

Mr. FETTUS. They wouldn't be approached like that, because, basically, you would have an entire reset where—we have right now a licensed consolidated interim storage facility in this country. You know that, right? In Utah. It exists. It is built, and it is licensed.

It will never accept a gram of waste, because Senator Orrin Hatch led the entire congressional delegation to put a wilderness around it in a circle to ensure waste will never be shipped there.

Mr. PETERS. Let me just suggest, what I am inferring is that there is not a lot of enthusiasm among the States to accept any amount of undefined—any undefined or defined amount of nuclear waste. There just isn't. You know, I have experience in local politics. No one even wants a house next to their house. I mean, to me, it is—you know, when you talk about federalism, the magic of federalism is the supremacy clause and the ability of the Federal Government to say: You know what? I have to—I can't—I have to look at all you squabblers, and I have to say, from a technical perspective, that this is a safer place to put this waste in this geology, per this engineering, per this licensing process, than say next to 80— you know, 8 million people on the coast, on a military base, that is a better—that this risk is lower.

I mean, I think, from a practical standpoint, I don't know where you get the idea that some State is going to come, you know, give you a high five for putting nuclear waste there. It is just not going to happen.

And it is up to this committee and the Federal Government to say: From a technical perspective, from a safety perspective, it is not going to be in these dangerous places; it is going to be in safer place.

That is why I support all three of these bills, and I think we got to move forward.

Mr. Chairman, I yield back.

Mr. TONKO. The gentleman yields back.

The Chair now recognizes Mr. Walden, full committee ranking member for five minutes to ask questions.

Mr. WALDEN. Thank you, Mr. Chairman. And thanks to all the witnesses. We have got a couple of hearings going on today, so I am bouncing back and forth.

Mr. Barrett, during a hearing before this subcommittee in May of 2015, Washington State Assistant Attorney General Andy Fitz testified that the Federal Government's inaction on the Yucca Mountain license application was harming DOE's obligations to clean up defense high-level waste at the Hanford site, which is right across the river from my district.

Could you speak to the importance of completing the licensing process as it relates to sites like Hanford?

Mr. BARRETT. Yes, sir.

The Federal repository, which Yucca Mountain is the site on the table right now, would take care of the commercial fuel but would also take care of all our defense high-level waste of which the primary source is at Hanford. So it would a solution for that.

Mr. WALDEN. And you know a bit about that the Hanford site, right?

Mr. BARRETT. Yes, sir.

Mr. WALDEN. And the leaking tanks?

Mr. BARRETT. Yes, sir.

Mr. WALDEN. Yes. They were designed to last, what, 40 years, 20 years?

Mr. BARRETT. Long ago.

Mr. WALDEN. Yes. Like World War II.

Mr. BARRETT. Yes, sir.

Mr. WALDEN. We have kind of exceeded their lifespan.

Mr. BARRETT. Well, precautions are taken to keep them safe. But, yes, sir.

Mr. WALDEN. Sure.

And the clean-up will take how long.

Mr. BARRETT. Long time.

Mr. WALDEN. Yes.

Like a century.

Mr. BARRETT. A long time, sir.

Mr. WALDEN. Yes. And, meanwhile, it sits there along the banks of the Columbia River. We have got leaks. We have got tunnels that are collapsing on top of old railcars that are somewhat radioactive. I mean, they are doing their best, but they are going to reprocess, and then where does that go?

Mr. BARRETT. It needs to go to a geological repository somewhere, like a Yucca Mountain facility.

And, yes, this generation has a responsibility for what we did 60, 70 years ago.

Mr. WALDEN. That is exactly right.

So, Ms. Korsnick and Mr. Barrett, H.R. 2699, the McNerney-Shimkus legislation provides accelerated interim storage that is integrated into a permanent storage program. It is authorized up to 50 million per year through general appropriations until the operation repository commences.

Do you think that is enough to initiate an interim program and an appropriate source of funding?

Ms. KORSNICK. I mean, I think it is a good start. The reality is we just need to get on with it, you know, get started. And there is plenty of money that has been set aside for this. And we need to move forward.

Mr. WALDEN. Mr. Barrett.

Mr. BARRETT. Yes, sir. I think it is a good start as well. I think it needs to be complimented with a banging of the heads together between the Federal Government and Nevada to compliment that with using the principles of the blue ribbon commission of empowerment to work the needs out for both.

Mr. WALDEN. I think I know a couple of my colleagues are more than willing to bang some heads together.

H.R. 3136, Ms. Matsui's bill, would allow appropriations for interim storage of the Nuclear Waste Fund for up to 25 percent of the interest on the fund, or upwards of $300 million a year. It seems to me that diverting funds from the waste fund for interim storage that is not linked to Yucca risks increasing the financial burden on ratepayers to pay for a permanent repository.

Is that a fair concern? Ms. Korsnick and Mr. Barrett?

Ms. KORSNICK. I think, you know, it is. But I guess I would qualify that. There has been so much money set aside. I mean, $41 billion, really, and 1.5 million in interest every year. It is almost hard to imagine that you can't build interim and a long-term repository for that amount of money.

Mr. WALDEN. But hasn't the $41 billion basically already been spent by the Congress in some other—no. OK.

Yes. Right.

Mr. Barrett.

Under a CBO issue we ran into——

Mr. BARRETT. I believe interim storage, you know, can be done. And I think it would not be a major issue as far as the funds that are set aside.

Accessing those funds that are set aside is more of a challenge that you have.

Mr. WALDEN. That is the issue.

So, I guess the point is Mr. Shimkus and others, but especially John, has really led the effort in Congress after Congress. And we reached a bipartisan agreement in the last Congress. We got 340 votes on the House floor to resolve this issue for the ratepayers, the taxpayers, and for the environment. And for the life of me, I can't understand—oh, that's right. The Constitution still gives Nevada two Senators every election.

We have got to solve this for the safety and security of the country, for the environment, for the ratepayers. This is nuts. And I think a lot of us on this committee understand that. Some of us have nuclear facilities right on fault lines, like in California and around, that we need to find a solution here.

And so, I just—I thank our witnesses for participating today. I remain fully committed, Mr. Chairman, to work with you or whoever or try and push this issue forward and get the solution we promised ratepayers many years ago.

Thank you. I yield back.

Mr. TONKO. You are welcome. The gentleman yields back.

The Chair now recognizes the gentleman from Florida, Mr. Soto, for five minutes, please.

Mr. SOTO. Thank you so much, Mr. Chairman.

I guess my first question for the panel is, is there any other alternative to storing it in a place like Yucca Mountain? For instance, further deep into the Earth's core or through other methods? Or is there pretty much the only one that is known to us right now. It would be great to start from left to right.

Ms. KORSNICK. Well, I think one of the biggest things to do is just look at the rest of the nuclear community in the other countries that have nuclear. And long-term repository is the solution of choice. Finland is probably the best example. They have licensed and are constructing a deep geological repository. But, you know, there are other countries, Sweden, Switzerland, France, others that are making notable progress.

So I think if you look at the broad nuclear community, deep geologic repository is the solution of choice.

Mr. SOTO. Mr. Halstead, do you think there is an alternative?

Mr. HALSTEAD. Well, first, I want to say the countries where success is occurring, as Ms. Korsnick said, are the very countries that have given up the approach we are using with Yucca Mountain. And, in fact, particularly, in Germany and France and the U.K., they started with forced siting, and it didn't work. So the first thing for finding a repository site, you have to have both good science, but you have to have the consent of the affected community.

Rather than talk more about repositories, I think there is some really interesting positive early results from the work on deep borehole disposal, and that is a whole another topic we could get into, but——

Mr. SOTO. How deep——

Mr. HALSTEAD [continuing]. Well, certainly, you are talking in some places about going beyond 2,000 meters. Pretty deep. It depends where the formation that you are seeking lies. But this is a very positive alternative. It also is a particularly attractive alternative for the waste that is at Hanford, some of which DOE now wants to reclassify as being less than high-level waste.

Now, there are pros and cons of that approach. But depending on how that decision works, there are certain types of waste at Hanford that might be a really good test for the deep borehole disposal approach.

Mr. SOTO. Thank you.

Mr. Keyser, is there a possibility of—is this what is really holding back some of the new nuclear facilities from going up? Because a lot of us want to get to renewable and clean energy by 2050 and believe that nuclear would be a part of that.

How much is this holding back that really becoming a reality?

Mr. KEYSER. I think it is the biggest force of communities, opposition to nuclear power, because they are going to be stuck with waste sitting on their sites at some point, 1982 is when this promise was made that this fuel would move off, 1998, it was supposed to be moved off. So we are almost 40 years past when Congress made a program, 20 years past when the stuff was supposed to be gone.

So it is now hard to talk to communities about developing new nuclear when the promises haven't been upkept in the past. We know they are safe. We know they are efficient. But we have to get to the point where we move that fuel to a national repository. Folks know they are not going to be stuck with that fuel. We can redevelop not just the new jobs in the energy implications of new nuclear facilities, which they are—fortunately, a couple reactors under construction now where we have thousands of members building them, but also redevelopment of those existing sites, whether they are for new nuclear or some other source of clean generation or sit in proximity to our Nation's grid making very efficient places for redevelopment, which could bring jobs back to those communities that were hurt so badly by the decommissioning of these nuclear plants.

Mr. SOTO. Thank you.

Mr. Barrett, we have a facility, Crystal River, that is closed now, two other facilities that are aging but could potentially be rehabbed.

What is the impact of a long-term—of having a closed reactor like Crystal River with rods just sitting there right now to the long-term environmental protection of Florida?

Mr. BARRETT. For safety purposes, the utilities will maintain those safe under NRC control. So there is no short-term safety issue with the fuel being stored there from our view.

I think in the long term, that is not where fuel needs to be, sitting on the seacoast right there at Crystal River and anywhere else. So I think that needs to be moved. It was never the agreement that we had. Those communities never consented to being a long-term storage. And those materials need to move somewhere that we need to have soon.

Mr. SOTO. Thank you.

And just one quick question. Mr. Fettus, can you—are these able to be recycled one day, you think, if we had the proper technology?

Mr. FETTUS. No.

Mr. SOTO. OK.

Mr. TONKO. The gentleman yields back.

The Chair now recognizes the gentleman from Ohio, Mr. Johnson, for five minutes.

Mr. JOHNSON. Thank you, Mr. Chairman.

Mr. Keyser, thank you for taking time today to testify at this important hearing and for your past work with the Ohio Valley Regional Development Commission. My staff and I have great relationships with John Hemmings and the OVRDC staff and enjoy working with them on economic development opportunities in southern Ohio. OVRDC's work along with the ARC, the Appalachian Regional Commission, and the other development districts are so vitally important to ensuring eastern and southeastern Ohio receives the support it needs to be an economic player in our State and in our region.

So we are happy to have you here today representing the International Brotherhood of Electrical Workers. So tell me, what role do IBEW workers play within the nuclear industry?

Mr. KEYSER. We are the largest union in the nuclear industry. We represent most of the workers in nuclear generation, so these plants are loaded with our members. Not only there but at the old enrichment facilities, at weapons grade facilities, at all these facilities I have heard named today. We have IBEW members onsite doing core work, whether it is in the operations or just the electrical construction or capital improvements or decommissioning of sites. So it is a huge workforce, about 15,000 members in generation alone.

And it is important to us to keep those jobs. So they, nuclear, out of all the utilities, employs more people per megawatt hour. So they are larger employers. These jobs pay dramatically higher wages because of the skill sets than do most jobs comparable in their communities. They are the anchors of these communities. As you know from the Portsmouth site and the workers there, without those wages in that community, where would the Portsmouth and Chillicothe and Jackson, Ohio——

Mr. JOHNSON. Sure.

And you sort of answered this a little bit, because I was going to ask you, these jobs are really good-paying jobs, are stable jobs. Explain that a little—they are stable because they are so long term, right? I mean, you are there for a long time once you are going an operation.

Mr. KEYSER. Right. They are stable. And in the utility space, in general, those jobs have the lowest turnover. So they are not as reflective as the marketplace. If the market takes a dip, you don't get rid of your generation workers, right? It doesn't happen. So they are immune to those sorts of downturns.

And in many places in our economy, during the Great Recession, without the utility bases in some of these communities, especially in rural areas, they would have been completely decimated with no opportunity for recovery. So they are good solid long-term jobs. These folks invest heavily either through our construction apprenticeships or through trainings in conjunction with our utility partners. And these are long-term careers. They aren't just jobs.

Mr. JOHNSON. OK.

All right. Now, I know your members are eager to get to work in expanding the country's nuclear power sector.

From your perspective, this is kind of some out-of-the-box thinking. What are the main barriers preventing the expansion of nuclear power?

Mr. KEYSER. I think it is public perception. And I think a lot of it comes with this argument that we are having today on the disposal of spent fuel. And since we are focused on that, I think that is maybe the number one reason currently is the handling of fuels. And we know that it is safe. We know that our workers are in these facilities constantly.

As I said in my opening statement, the high watermark for industrial safety is at these nuclear facilities. We have very few occupational injuries in general. I don't know of a single death that has ever occurred from radiation exposure from one of our workers in these facilities.

So a very high skilled, very heavily regulated industry, very safe jobs.

Mr. JOHNSON. OK. And you touched on this a little bit in a previous answer, and you just alluded to it here as well. So do you think the proposed legislation we are discussing today would help advance the nuclear industry and benefit your members? Do you want to add anything to that?

Mr. KEYSER. Yes, absolutely I do.

You know, I think an interim facility is okay. And it maybe should happen to go ahead and start moving fuel. But it can't happen, and nobody is going to opt into it voluntarily without the back-drop of a permanent facility going to come online where people can move this stuff.

So, if we are asking people to bid on interim facilities at multiple locations where they have no guarantee that that fuel is not going to be then moved away from them and into a national repository, why would they bid on that any more than Nevada would bid on Yucca Mountain?

So we do have—to Congressman Peters' point, we do have an obligation here as a Nation to handle these fuels and put them in the safest place possible and move those and keep this—keep the nuclear industry, which is, you know, around-the-clock base-load power generation, especially as we move to more renewables. We have to have nuclear as a major part the generation mix.

Mr. JOHNSON. OK. I am going to give you four seconds back, Mr. Chairman.

I yield back.

Mr. TONKO. Thank you very much. The gentleman yields back.

I am told they may call votes within five minutes, so we are going to ask everyone to keep their questioning to three minutes.

We now recognize the gentlewoman from California, Ms. Matsui.

Ms. MATSUI. Thank you, Mr. Chairman.

Thank you for calling this hearing and considering my proposal, the STORE Nuclear Fuel Act.

Now, I think you know, my situation, I have a decommissioned plant in Sacramento, SMUD. And we would really like to ensure that we can remove forward on this. And, you know, this repository program is still stalled, and we have this potential for consolidated interim storage. And it is critically important this now if we want to be serious about getting something done.

And I want to be clear: My bill is not intended to act as a substitute for existing law; rather, it is an addition so that we can make some progress here. And I think that is really very important. Until we really get to the Yucca, or whatever we are talking about, we have got to make some progress here.

Mr. Barrett, I know you have said before that it would be wise to reengage in good-faith discussions with the State of Nevada in an effort come to some sort of solution that is agreeable to all parties. But a process like this requires time and patience on all sides. When you say, "reengage to come up with a win-win," what do you have in mind? What would be the omnis, and how do we go about it?

Mr. BARRETT. Well, I believe if the Federal Government engages with Nevada following the principles of the blue ribbon commission of empowerment, we can reach an agreement that would advance where we are picking up some of the points that Mr. Fettus said as part of an agreement between those two parties. So I believe a win-win can be done.

Ms. MATSUI. Well, the problem is, though, is that this new relationship we are talking about with Nevada, I mean, I am just looking at this.

Mr. Halstead, can you give me some sense of how long you believe the licensing process at NRC could take and whether there are steps that the State of Nevada will take to further delay construction of a repository or an issuance of a necessary license to DOE to actually possess and begin moving used fuel to the site?

Mr. HALSTEAD. Well, one of the questions that DOE needs to answer is exactly what licensing would cost if restarted. I think it is foolish both to force a licensing restart or appropriate funds until you know the answer. Now, we have estimated it, based on DOE, NRC, our own knowledge, at about $2 billion over 4 to 5 years.

And let me correct what Mr. Shimkus says. We are actually spending Nevada's money on this. Since 2010, we have spent about $26 million in State money. Last Federal money we got was in 2009. The State—I have told the legislature: Don't expect any Federal money.

And the legislature has said: Tell us how much money you need.

This is how serious the opposition is.

So when we say 4 to 5 years and 2 billion, that might be a low number because there are so many parties to the licensing proceeding. And, remember, the Timbisha Shoshone Indian Tribe and the Native Community Action Council, two Native American groups, they have not weighed in yet.

Ms. MATSUI. OK. I know I have gone beyond my time, but I just want to say that I don't believe it is fair for my constituents to have to wait through this process either.

So I yield back my time.

Mr. TONKO. The gentlewoman yields back.

The Chair recognizes the gentleman from Missouri, Mr. Long, for three minutes.

Mr. LONG. Thank you, Mr. Chairman.

Mr. Barrett, Congress is no doubt to blame for the failure to fund the licensing process. But it seems that there has been a huge gap between the Department of Energy and the State of Nevada and that they are operating using two different sets of facts and don't trust each other, which is leading, of course, to further delays.

In your testimony, you recommend the Department of Energy should increase engagement with Nevada to move this forward. In your heart of hearts, could or would that really work to move this process along?

Mr. BARRETT. I think it would be better if it was another agency or some other Federal organization. I tried to use the word Federal Government to engage with Nevada, because I think there is an awful lot of old baggage with the Department of Energy, and we could have a better engagement with a different type of organization.

But the law is the law, and it is the DOE at the time. But I believe there are many better arrangements.

Mr. LONG. I will take that as a no.

In order to build trust between the Department of Energy and the State of Nevada, would one step be to complete the safety license process for Yucca so every stakeholder will be on the same page about the operational safety and feasibility of the Yucca project?

Mr. BARRETT. Yes, sir.

Mr. LONG. I will take that as a yes.

And there would be time to work out an agreement between the licensing and the operating permit, correct?

Mr. BARRETT. That would be my goal.

Mr. LONG. Switching gears here, Mr. Barrett.

One of the best ways to reduce carbon emissions is to research and develop innovative technologies that improve our current methods of electricity generation. There are excited developments in the nuclear industry with small-scale reactors that are being built as we speak.

Does the lack of progress we have made on long-term nuclear waste disposal policy hinder or discourage innovation in this field of a new type of nuclear reactors?

Mr. BARRETT. Unfortunately, yes.

Mr. LONG. OK. I am an auctioneer. I did my three minutes in two. I hope everyone appreciates that.

I yield back.

Mr. TONKO. The gentleman yields back.

The Chair now recognizes the Representative from Michigan, Congresswoman Dingell.

Mrs. DINGELL. Thank you, Mr. Chair.

And I love the gentleman from Missouri.

We have established the fact that all of us are really worried about what is happening. The waste is just currently sitting at active and decommissioned nuclear plants across the country. Michigan is one of them. So I want to focus my questions on stored nuclear fuel in the Great Lakes region and the dangers it presents. And if you don't know the answer, just tell me, because we are quick on time.

Does anyone on this panel know how many tons of stored nuclear fuel exists in temporary storage locations across the shores of the Great Lake regions?

Mr. FETTUS. I think it is 10,000 metric tons in Michigan and a bunch more in Canada.

Mrs. DINGELL. 60,000, yes.

Mr. FETTUS. OK.

Mrs. DINGELL. So it is about 70. I am sorry. It is 50,000 in Canada.

So, in total, there are over 60,000 tons of highly radioactive spent nuclear fuel surrounding the Great Lakes. That is almost as much as all the spent nuclear fuel in the United States, which is nearly 70,000 tons.

Should we prioritize the removal of spent nuclear fuel away from the Great Lakes?

Any of you want to answer that?

Mr. BARRETT. Yes.

Mr. HALSTEAD. I think the first thing you want to do is move it from wet storage to dry storage, regardless of where it is. That is a really big concern at the lakeside plants.

Mr. FETTUS. Agreed.

Mrs. DINGELL. What would happen if we were to have a leak into the Great Lakes?

Ms. KORSNICK. I guess if I could just address that in general.

I mean, where it is stored right now, it is stored very safely. This fuel is not liquid. It is not like it leaks. These are metal rods that are put in stainless steel casks that are stored on concrete pads. So it is not like it can sort of spring a leak, if you will. They are seismically designed and designed for a variety of hazards.

The NRC has very close scrutiny over any of this used fuel that is stored. It is stored very safely at all the sites today.

Mrs. DINGELL. Do other of you agree with that?

Do you think we should be storing nuclear waste in the Great Lakes? 20 percent of our fresh water?

Mr. BARRETT. I agree it is being stored safely, but it doesn't belong there for hundreds of years.

Mr. FETTUS. I think we are trying to chart a way forward that can actually work and move it and faster than the road we are on now.

Mrs. DINGELL. Anybody else want to comment?

Mr. KEYSER. I would just agree that it shouldn't be stored there forever, but it is stored safely. We have workers in those facilities and around those facilities. They feel safe moving it. We have never had an accident. We have never had major radiation exposure to any of our members, so—but, again, these——

Mrs. DINGELL. We always say that until it happens.

Mr. KEYSER. Right. These aren't designed to last forever, so, yes, they need to be moved.

Mrs. DINGELL. I yield back my time. I am not as good as the gentleman from Missouri, but——

Mr. TONKO. The gentlewoman yields back.

The Chair recognizes Congressman Flores for three minutes, please.

Mr. FLORES. Thank you, Chairman Tonko and leadership for holding today's hearing.

Thank you for the witnesses being here today.

I want to talk about the cost of starting over. The country lost a substantial amount of time that was very valuable and billions of dollars, tens of billions of dollars, of hardworking taxpayer money after the previous administration shut this program down.

Now opponents say that Yucca Mountain will be too time-consuming or costly to resume the licensing process.

Ms. Korsnick, do you believe it would be cheaper and quicker to complete the Yucca licensing process or to start over with a consent-based licensing process?

Ms. KORSNICK. I think it would be cheaper to go forward with the Yucca Mountain process.

Mr. FLORES. OK. I am also concerned that if we don't solve the storage problem, we won't be able to deploy the new technologies that Mr. Long was talking about that are zero-based emissions technologies.

Would you also agree, Ms. Korsnick?

Mr. Barrett, based on your experience and observations at the Department, would you expect it to be significantly more effective for the taxpayers and quicker to move forward with the review of the Yucca license or starting over?

Mr. BARRETT. Yes, I do.

Mr. FLORES. OK. Mr. Barrett, where would the funds come from to start a second repository process?

Mr. BARRETT. I believe they could come from the Nuclear Waste Fund. The original program did. I believe the Yucca Mountain should go forward, and we should supplement that with another search, maybe along the lines of Mr. Fettus', but we need to move forward.

Mr. FLORES. Ms. Korsnick, in that case, what would the—you know, we have got a shrinking nuclear fleet, unfortunately. What would the cost be if we had a new fee structure, a new revenue structure on top of the existing structure?

Ms. KORSNICK. Yes. So that is a significant challenge. The current plans are very stressed, as you know in the marketplace. And so to restart a nuclear waste fee on top of the challenges that they are facing today is very significant.

Mr. FLORES. OK. It would make it less economic. It would probably make it less attractive to invest in the new technologies we need to have a zero emissions economy, right?

Ms. KORSNICK. That is correct.

Mr. FLORES. OK. Mr. Barrett, what would the cost be for site characterization and related design work in starting the whole NRCC process over?

Mr. BARRETT. Approximately $10 billion or so.

Mr. FLORES. OK.

All right. I yield back the balance of my time.

Thank you.

Mr. TONKO. The gentleman yields back.

The Chair now recognizes Mr. McNerney, for three minutes.

Mr. MCNERNEY. Well, I thank the chairman. And I want to thank my colleague Mr. Shimkus for hard, persistent work on this very difficult subject.

Ms. Korsnick, would you briefly discuss again or reiterate again the need for permanent and interim storage to be considered part and parcel for a solution?

Ms. KORSNICK. Yes. So the long-term answer is the long-term geologic repository. And, you know, I would also just remind the committee that we are talking about these sites that you are putting the current fuel in. There is still 95 percent good energy in these used fuels. It is really better characterized as future nuclear fuel. So you want to put it someplace where you can get it back, because in the future you are going to want it back.

Mr. MCNERNEY. So it is recyclable in your opinion.

Ms. KORSNICK. That is correct.

You are going to want to use fuel that is in there in a different type of reactor.

And so the current place that we want to have it is a long-term repository as has been mentioned here a few times. Nobody wants to sign up for interim storage if they don't appreciate that there is a long-term repository.

But the interim repository can consolidate the fuel into fewer locations on its way to the long-term repository.

Mr. MCNERNEY. Thank you.

Mr. Barrett, communities clearly need to be involved in the process. But as you have pointed out, I think just about every State that has had a location considered has blocked progress on that.

So what kind of framework could we use that would work in terms of getting enough buy in in the State to move forward?

Mr. BARRETT. I think if both sides—the Federal side and the individual States engage with the principles of the blue ribbon commission of empowerment and working together, you can work out arrangements that can meet the needs of both, safety wise, clearly, but also economically. In the development of the State, things can be worked out if they can get together and get away from the emotion that has dragged us down in the past.

Mr. MCNERNEY. I mean, I understand that Sweden has a pretty widely accepted nuclear waste repository program. Is that right?

Mr. BARRETT. Yes, sir, it does. But there is a big difference between Sweden and Finland and others is they have no State level of government. If we had no State level of government here, we would have had this solved decades ago. But we have a thing called States in the United States of America.

Mr. FETTUS. I firmly agree with that.

Mr. MCNERNEY. I would like to ask Mr. Fettus, if I have time.

Your proposal of using RCRA and amending the Atomic Energy Act so radioactive materials comply with environmental laws would open up a whole can of legal issues.

Mr. FETTUS. I think that can of legal issues is already open. I am trying to put a cap on it.

Mr. MCNERNEY. I mean, I think that is a debatable whether it would make it worse or better then.

Mr. FETTUS. Look at the website in New Mexico, Congressman. To the extent that there is any acceptance, it is the State RCRA authority.

Mr. MCNERNEY. I yield back.

Mr. TONKO. OK. The gentleman yields back.

The Chair recognizes the gentleman from Georgia for three minutes.

Mr. CARTER. Thank you, Mr. Chairman.

Thank all of you for being here. This is very important subject, obviously for us in the State of Georgia. Guess which State is the only State that has two nuclear reactors under construction right now. It is the State of Georgia. Not only is it—not only is that important, but it is also important because it means that Georgians will be able to pull more energy from a stable and a clean source, such as nuclear power.

And, you know, I get confused and frustrated sometimes why there is so much misinformation out there about nuclear power. It is clean energy. And, yes, we do need to address the issue of the depositories and repositories and what we do with the waste. But, you know, we have got that solution in hand. We are just not taking advantage of it.

I want to start with you, Ms. Korsnick, and ask you, how often is nuclear waste actually transported across the country?

Ms. KORSNICK. It is actually transported fairly regularly. There has been probably 1,300 used fuel shipments across the United States over the last few decades, all done very safely.

Mr. CARTER. Can you give me some ideas of the precautions that are taken? I suspect that they are numerous.

Ms. KORSNICK. Yes. If you just imagine the package to begin with, so the fueler of these long metal rods, and they are placed in a cask. The cask for every 1 ton of fuel that you have, there is probably 5 to 7 tons of cask, if you will, that are protectively wrapped around, if you will, this fuel. These casks are designed and licensed by the Federal Government. There have been all kinds of different postulated accidents of which they are all designed against. So it is all done very safely.

Mr. CARTER. It is done safely, and it is done quite often. So it is not as if this would be something that is new to us if we were to a place like Yucca Mountain to use as a permanent site, if you will.

Ms. KORSNICK. That is correct.

Mr. CARTER. It appears to me that several of these bills address only the interim side and not the long-term issue that is out there. How important is it—I know it is important that we obviously concentrate and focus on the interim side, but how important is it that we also look at a long-term solution?

Ms. KORSNICK. Well, it is critical. As we mentioned, you know, earlier, if you don't have the long-term repository, when you sign up for the interim repository, you might feel like you are signing up for the long-term repository. And so it really negates the idea of having a consolidated interim storage until people have a view that there is a long-term answer.

Mr. CARTER. Right.

And just one last thing. And that is, Mr. Keyser, you mentioned about the jobs. And those are very important. Let me tell you, the construction of the plant too, the jobs that are being supported by that are just enormous.

Mr. KEYSER. Yes. Absolutely. We have over 4,000 building trades members, not just IBEW on that site every day. So massive project. And not only that, but the advanced manufacturing of some of the components of the IBEW——

Mr. CARTER. Thank you.

And I yield back.

Mr. TONKO. The gentleman yields back.

The Chair recognizes very patient gentleman from North Carolina—South Carolina, Mr. Duncan, for three minutes.

Mr. DUNCAN. I was going to correct you, Mr. Chairman, but South Carolina.

Thank you so much. Thanks for being here. I sat patiently, as you said.

Let me just make a few points because a lot of questions have been asked. But the high water mark of Yucca Mountain is when the law was passed by the United States Congress for a long-term repository. Construction started, but it started the downhill spiral of folks saying no, even though we did a lot of study of a lot of different sites around the country and ultimately landed on Yucca Mountain as the ideal repository for long-term storage for nuclear waste.

And I hear a lot of talk about an interim storage site. Well, I can tell you, nobody wants to be the interim storage site, I think Ms. Korsnick said it best, because they don't believe and trust the Federal Government will ever get to a long-term repository.

Ratepayers in this country were sold a bill of goods. Nuclear waste sits at commercial sites at 121 different locations around the country, on the shores of Lake Erie and Ohio and south on the shores of Lake Keowee in South Carolina, on the shores of the Savannah River in Georgia. And we know that nuclear waste, and we have had leaks in Illinois, six different sites in Illinois.

So the American people and the States don't trust Federal Government to ever get to this point. Yucca Mountain was chosen. I have been there. I stood on top of that mountain, and I said: If we can't put nuclear waste here, we will never put it anywhere else, because we won't pick another site. There will be too much politics involved.

Back when the law was passed, and I use the word "law" because it is the law of the land to put the nuclear waste from commercial reactors and defense waste at Yucca Mountain. I say that the taxpayer, the ratepayers were sold a bill of goods. I used a number earlier. I misspoke. South Carolina ratepayers have paid $3 billion for the operation and construction of Yucca Mountain. That is $3 billion of not taxpayer money; that is ratepayers and their utility bills, pennies at a time. But South Carolinians have paid $3 billion for this.

We have got nothing for it. And I say that no one wants to trust the Federal Government with regard to doing what they say they are going to do. They brought a bunch plutonium to the Savannah River site and said: We are going to push for MOX, mixed oxide fuel, and we are going to convert that plutonium from the non-proliferation agreements into useable fuel. Guess what, folks? MOX is mothballed.

And that plutonium still sits on a concrete pad in a metal building at Savannah River site because we are not going to convert it to MOX. That is another example where the Federal Government has lied to the American people. That plutonium can go to Yucca Mountain as well.

I think Ms. Korsnick said: We need to be able to access that. As technology gets there, we can reprocess this fuel and reuse it. Absolutely. Yucca Mountain has ingress and egress. So we can take the waste and we can remove the waste to use it.

Folks, it is the law of the land. This is time to move forward. And let's let the ratepayers of South Carolina and other States get what they are paying for, what they have paid for.

Mr. Chairman, thanks for your leadership on this. I know we have a vote.

I yield back.

Mr. TONKO. OK. Thank you. The gentleman yields back.

I thank all of our witnesses for joining us today for your input, which is very critical to a very, very critical issue.

I remind Members that, pursuant to committee rules, they have 10 business days by which to submit additional questions for the record to be answered by our witnesses. I would ask each witness to respond promptly to any such questions that you may receive.

I also have here a series of documents approved by both sides. I request unanimous consent to enter these documents into the record.

Without objection, so ordered.

Mr. TONKO. And, with that, at this time the subcommittee is adjourned.

[Whereupon, at 11:40 a.m., the subcommittee was adjourned.] [Material submitted for inclusion in the record follows:]

Prepared Statement of Hon. Greg Walden

The Energy and Commerce Committee has an enduring and strong bipartisan record supporting nuclear energy. Nuclear is a critical component of our nation's energy system. It also has been vital to our national security, powering the nuclear navy and providing for our common defense.

Not only is nuclear power an affordable and reliable energy source, it's also emissions-free. Any serious efforts to reduce emissions from energy production to address the effects of climate change must include the continued use and expansion of nuclear power. And there is great potential—if we get the policies right--to benefit from nuclear energy far into the future.

Over the past few years we've moved legislation to lay the groundwork for advanced nuclear and to ensure more efficient regulation of the existing reactor fleet.

We've explored policies that will ensure a nuclear infrastructure for tomorrow— ranging from advanced small modular reactors like those under development by Oregon-based NuScale, currently in NRC licensing, to advanced fuel systems for the next generation of reactors.

Yet, as we look forward, we have responsibility to ensure we implement the existing policies that address the issue of long-term storage of spent nuclear fuel, and the defense legacy waste that the federal government has a responsibility for cleaning up.

This is no small matter. 35 years ago, Congress enacted the Nuclear Waste Policy Act into law. This law was the culmination of decades of experience by the federal government to develop a policy to permanently dispose of high-level radioactive waste and commercial spent nuclear fuel.

Some of the material was created during the Manhattan Project and through the Cold War at the Hanford site, a vital national security facility located on the Columbia River about 40 miles north of my district. Today, this nuclear material sits on a vibrant waterway waiting to be processed and transported to the Yucca Mountain repository in the Nevada desert.

The Nuclear Waste Policy Act also established a fee tied to the generation of nuclear energy to finance the costs of a multi-generational disposal program. Along with 33 other states, Oregon ratepayers fulfilled their financial obligations under the law and paid the Department of Energy over $160 million to dispose of commercial spent nuclear fuel.

I've noted in the past how the Trojan nuclear power plant, located in northwest Oregon, stopped producing electricity in 1993, with the expectation that DOE would begin to remove the spent fuel in 1998, as was required by law. That never happened and since the plant's decommissioning activities were completed in 2007, only spent nuclear fuel remains stranded at the site, hampering any redevelopment efforts surrounding it.

This example is repeated across the nation, with states and communities waiting for DOE to fulfill its obligations and dispose of the spent fuel.

As we all know, the Federal Government has been prevented from completing the licensing process for a permanent repository. The costs to the American taxpayer to pay for the federal government's delay in opening the Yucca Mountain repository have more than doubled to $35 billion since 2009 and that figure continues to escalate rapidly as time goes on. Meanwhile, the federal government is paying out nearly a billion dollars a year from the judgement fund for its failure to dispose of the waste.

Against this backdrop, Mr. Chairman, I appreciate your moving forward on examining legislative reforms that can help to restart this process.

The Energy and Commerce Committee should continue to lead the effort to ensure the Federal government meets is moral and fiduciary responsibility to clean up its defense waste and ensure the permanent, safe disposal of spent nuclear fuel.

We made important strides in the last Congress to reform the fundamental statute to help to accelerate this complicated process. My friend and the Republican Leader of this subcommittee, John Shimkus, led the work in the House to pass the Nuclear Waste Policy Amendments Act by a vote of 340–72. Unfortunately, that effort fell short in the Senate.

But we know from the last Congress, and from the strong bipartisan support both on this Committee and in the House of that legislation, how a thoughtful and deliberate legislative process produces good policy.

I'm pleased to see this past work has informed the current work, particularly in HR 2699, led by Mr. McNerney, which follows closely the H.R. 3053 from the last Congress.

This bill provides for accelerating interim storage of waste without undermining the important system for permanent disposal established in the underlying law. This represents the best path forward for getting the nation to a licensing decision, which is necessary for public confidence in our nuclear waste program, no matter the outcome of that decision.

Thank you, Mr. Chairman for taking the lead on this important issue.

Prepared Statement of Hon. Doris O. Matsui

Thank you. Chairman Pallone.

Finding a solution to managing the disposal of spent nuclear fuel has been a top priority of mine for years . . . particularly as my district's utility, the Sacramento Municipal Utility District, is one of the many across the country forced to play host to this dangerous radioactive material long after they committed to do so.

I think we all agree that this stalemate is unsustainable. The best and most pragmatic path forward involves a consolidated interim storage program that will engage with affected states and local governments through a consent-based process.

That is why I have introduced the STORE Nuclear Fuel Act . . . which puts forward a plan that has historically garnered broad support.

The Federal Government has reneged on its promise to our constituents . . . a consolidated interim storage approach will allow the over 120 communities across the country to redevelop nuclear reactor sites that for many have been decommissioned for years.

I believe this is one of the greatest energy challenges of our time. We cannot afford to play politics any longer. I am grateful to the Committee for bringing my bill up for discussion today.

Thank you and I yield back.

116th Congress, 1st Session, H. R. 2699, to Amend the Nuclear Waste Policy Act of 1982, and for Other Purposes, in the House of Representatives, May 14, 2019

Mr. MCNERNEY (for himself, Mr. SHIMKUS, Mr. PETERS, Mr. DUNCAN, Mr. CARBAJAL, Mrs. LESKO, Ms. BLUNT ROCHESTER, Mr. UPTON, Mr. KEATING, Mr. ALLEN, Mr. MICHAEL F. DOYLE of Pennsylvania, Mr. WILSON of South Carolina, Mr. COURTNEY, and Mr. BALDERSON) introduced the following bill; which was referred to the Committee on Energy and Commerce, and in addition to the Committees on Natural Resources, Armed Services, the Budget, and Rules, for a period to be subsequently determined by the Speaker, in each case for consideration of such provisions as fall within the jurisdiction of the committee concerned.

A Bill

To amend the Nuclear Waste Policy Act of 1982, and for other purposes.

Be it enacted by the Senate and House of Representatives of the United States of America in Congress assembled,

SECTION 1. SHORT TITLE; TABLE OF CONTENTS.

(a) SHORT TITLE.—This Act may be cited as the "Nuclear Waste Policy Amendments Act of 2019".

(b) TABLE OF CONTENTS.—The table of contents for this Act is as follows:

Title I—Monitored Retrievable Storage

Sec. 101. Monitored Retrievable Storage

(a) PROPOSAL.—Section 141(b) of the Nuclear Waste Policy Act of 1982 (42 U.S.C. 10161(b)) is amended—

(1) in paragraph (1)—

(A) by striking "1985" and inserting "2019"; and

(B) by striking "the construction of";

(2) in paragraph (2)—

(A) by amending subparagraph (C) to read as follows:

"(C) designs, specifications, and cost estimates sufficient to—

"(i) solicit bids for the construction of one or more such facilities; and

"(ii) enable completion and operation of such a facility as soon as practicable;"

(B) in subparagraph (D), by striking "this Act." and inserting "this Act; and"; and

(C) by adding at the end the following:

"(E) options to enter into MRS agreements with respect to one or more monitored retrievable storage facilities."; and

(3) by amending paragraph (4) to read as follows:

"(4) The Secretary shall, not later than 90 days after the date of enactment of the Nuclear Waste Policy Amendments Act of 2019, publish a request for information to help the Secretary evaluate options for the Secretary to enter into MRS agreements with respect to one or more monitored retrievable storage facilities.".

(b) ADDITIONAL AMENDMENTS.—

(1) IN GENERAL.—Section 141 of the Nuclear Waste Policy Act of 1982 (42 U.S.C. 10161) is further amended—

(A) in subsection (c)(2)—

(i) by striking "If the Congress" and all that follows through "monitored retrievable storage facility, the" and inserting "The"; and

(ii) by striking "construction of such facility" and inserting "construction of a monitored retrievable storage facility"; and

(B) by striking subsections (d) through (h).

(2) DEFINITIONS.—Section 2 of the Nuclear Waste Policy Act of 1982 (42 U.S.C. 10101) is amended—

(A) in paragraph (34), by striking "the storage facility" and inserting "a storage facility"; and

(B) by adding at the end the following:

"(35) The term 'MRS agreement' means a cooperative agreement, contract, or other mechanism that the Secretary considers appropriate to support the storage of Department-owned civilian waste in one or more monitored retrievable storage facilities as authorized under section 142(b)(2).

"(36) The term 'Department-owned civilian waste' means high-level radioactive waste, or spent nuclear fuel, resulting from civilian nuclear activities, to which the Department holds title."

(3) TECHNICAL AMENDMENTS.—Section 146 of the Nuclear Waste Policy Act of 1982 (42 U.S.C. 10166) is amended—

(A) in subsection (a), by striking "such subsection" and inserting "subsection (f) of such section"; and

(B) in subsection (b), by striking "this subsection" and inserting "this section".

Sec. 102. Authorization and Priority

Section 142 of the Nuclear Waste Policy Act of 1982 (42 U.S.C. 10162) is amended by striking subsection (b) and inserting the following:

"(b) AUTHORIZATION.—Subject to the requirements of this subtitle, the Secretary is authorized to—

"(1) site, construct, and operate one or more monitored retrievable storage facilities; and

"(2) store, pursuant to an MRS agreement, Department-owned civilian waste at a monitored retrievable storage facility for which a non-Federal entity holds a license described in section 143(1).

"(c) PRIORITY.—

"(1) IN GENERAL.—Except as provided in paragraph (2), the Secretary shall prioritize storage of Department-owned civilian waste at a monitored retrievable storage facility authorized under subsection (b)(2).

"(2) EXCEPTION.—

"(A) DETERMINATION.—Paragraph (1) shall not apply if the Secretary determines that it will be faster and less expensive to site, construct, and operate a facility authorized under subsection (b)(1), in comparison to a facility authorized under subsection (b)(2).

"(B) NOTIFICATION.—Not later than 30 days after the Secretary makes a determination described in subparagraph (A), the Secretary shall submit to Congress written notification of such determination."

Sec. 103. Conditions for Mrs Agreements

(a) AMENDMENT.—Section 143 of the Nuclear Waste Policy Act of 1982 (42 U.S.C. 10163) is amended to read as follows:

"Sec. 143. Conditions for Mrs Agreements

"(a) IN GENERAL.—The Secretary may not enter

into an MRS agreement under section 142(b)(2) unless—

"(1) the monitored retrievable storage facility with respect to which the MRS agreement applies has been licensed by the Commission under the Atomic Energy Act of 1954 (42 U.S.C. 2011 et seq.);

"(2) the non-Federal entity that is a party to the MRS agreement has approval to store Department-owned civilian waste at such facility from each of—

"(A) the Governor of the State in which the facility is located;

"(B) any unit of general local government with jurisdiction over the area in which the facility is located; and

"(C) any affected Indian tribe;

"(3) except as provided in subsection (b), the Commission has issued a final repository decision;

and

"(4) the MRS agreement provides that the quantity of high-level radioactive waste and spent nuclear fuel at the site of the facility at any one time will not exceed the limits described in section 148(d)(3) and (4).

"(b) INITIAL AGREEMENT.—

"(1) AUTHORIZATION.—The Secretary may enter into one MRS agreement under section 142(b)(2) before the Commission has issued a final repository decision.

"(2) FUNDING.—There are authorized to be appropriated to carry out this subsection—

"(A) for each of fiscal years 2020 through 2022, the greater of—

"(i) $50,000,000; or

"(ii) the amount that is equal to 10 percent of the amounts appropriated from the Waste Fund in that fiscal year; and

"(B) for each of fiscal years 2023 through 2025, the amount that is equal to 10 percent of the amounts appropriated from the Waste Fund in that fiscal year.

"(3) PRIORITY.—

"(A) IN GENERAL.—An MRS agreement entered into pursuant to paragraph (1) shall, to the extent allowable under this Act (including under the terms of the standard contract established in section 961.11 of title 10, Code of Federal Regulations), provide for prioritization of the storage of Department-owned civilian waste that originated from any facility that—

"(i) has ceased commercial operation;

and

"(ii) is located in—

"(I) an area that is of high seismicity; and

"(II) close proximity to a major body of water.

"(B) NO EFFECT ON STANDARD CONTRACT.—Nothing in subparagraph (A) shall be construed to amend or otherwise alter the standard contract established in section 961.11 of title 10, Code of Federal Regulations.

"(4) CONDITIONS.—

"(A) NO STORAGE.—Except as provided in subparagraph (B), the Secretary may not store any Department-owned civilian waste at the initial MRS facility until the Commission has issued a final repository decision.

"(B) EXCEPTION.—

"(i) FINDING.—The Secretary may make a finding that a final repository decision is imminent, which finding shall be updated not less often than quarterly until the date on which the Commission issues a final repository decision.

"(ii) STORAGE.—If the Secretary makes a finding under clause (i), the Secretary may store Department-owned civilian waste at the initial MRS facility in accordance with this section.

"(iii) NOTICE.—Not later than 7 days after the Secretary makes or updates a finding under clause (i), the Secretary shall submit to Congress written notification of such finding.

"(iv) REPORTING.—In addition to the requirements of section 114(c), if the Secretary makes a finding under clause (i), the Secretary shall submit to Congress the report described in such section 114(c) not later than 1 month after the Secretary makes such finding and monthly thereafter until the date on which the Commission issues a final repository decision.

"(C) NO EFFECT ON FEDERAL DISPOSAL POLICY.—Nothing in this subsection affects the Federal responsibility for the disposal of high-level radioactive waste and spent nuclear fuel, or the definite Federal policy with regard to the disposal of such waste and spent fuel, established under subtitle A, as described in section 111(b).

"(c) DEFINITIONS.—For purposes of this section:

"(1) FINAL REPOSITORY DECISION.—The term 'final repository decision' means a final decision approving or disapproving the issuance of a construction authorization for a repository under section 114(d)(1).

"(2) INITIAL MRS FACILITY.—The term 'initial MRS facility' means the monitored retrievable storage facility with respect to which an MRS agreement is entered into pursuant to subsection (b)(1).".

(b) CONFORMING AMENDMENT.—The item relating to section 143 in the table of contents for the Nuclear Waste Policy Act of 1982 is amended to read as follows: "Sec. 143. Conditions for MRS agreements.".

Sec. 104. Survey

Section 144 of the Nuclear Waste Policy Act of 1982 6 (42 U.S.C. 10164) is amended—

(1) by striking "After the MRS Commission submits its report to the Congress under section 143, the" and inserting "(a) IN GENERAL.—The";

(2) in the matter preceding paragraph (1), by striking "for a monitored retrievable storage facility" and inserting "for any monitored retrievable storage facility authorized under section 142";

(3) in paragraph (6), by striking "; and" and inserting a semicolon;

(4) in paragraph (7), by striking the period at the end and inserting "; and"; and

(5) by adding after paragraph (7) the following:

"(8) be acceptable to State authorities, affected units of local government, and affected Indian tribes.

"(b) REQUEST FOR PROPOSALS.—The Secretary shall issue a request for proposals for an MRS agreement authorized under section 142(b)(2) before conducting a survey and evaluation under subsection (a), and shall consider any proposals received in response to such request in making the evaluation.".

Sec. 105. Site Selection

Section 145 of the Nuclear Waste Policy Act of 1982 5 (42 U.S.C. 10165) is amended—

(1) in subsection (a)—

(A) by striking "select the site evaluated" and inserting "select a site evaluated";

(B) by striking "the most"; and

(C) by inserting "authorized under section 142(b)(1)" after "monitored retrievable storage facility"; and

(2) by striking subsection (g).

Sec. 106. Benefits Agreement

Section 147 of the Nuclear Waste Policy Act of 1982 (42 U.S.C. 10167) is amended—

(1) by inserting "the Secretary intends to construct and operate under section 142(b)(1)" after "storage facility"; and (2) by inserting "or once a non-Federal entity enters into an MRS agreement under section 22 142(b)(2)," after "section 145,".

Sec. 107. Licensing

(a) REVIEW OF LICENSE APPLICATION.—Section 148(c) of the Nuclear Waste Policy Act of 1982 (42 U.S.C. 10168(c)) is amended by striking "section 142(b)" and inserting "section 142(b)(1)".

(b) LICENSING CONDITIONS.—Section 148(d) of the Nuclear Waste Policy Act of 1982 (42 U.S.C. 10168(d)) is amended—

(1) in paragraph (1), by striking "has issued a license for the construction of a repository under section 115(d)" and inserting "has issued a final decision approving or disapproving the issuance of a construction authorization for a repository under section 114(d)(1)"; and

(2) in paragraph (2), by striking "or construction of the repository ceases".

Sec. 108. Financial Assistance

Section 149 of the Nuclear Waste Policy Act of 1982 is amended by inserting "authorized under section 142(b)(1)" after "a monitored retrievable storage facility".

Title II—Permanent Repository

Sec. 201. Land Withdrawal, Jurisdiction, and Reservation

(a) LAND WITHDRAWAL, JURISDICTION, AND RESERVATION.—

(1) LAND WITHDRAWAL.—Subject to valid existing rights and except as provided otherwise in this section, the lands described in subsection (c) are withdrawn permanently from all forms of entry, appropriation, and disposal under the public land laws, including the mineral leasing laws, the geothermal leasing laws, and the mining laws.

(2) JURISDICTION.—Except as otherwise provided in this section, jurisdiction over the withdrawal is vested in the Secretary. There are transferred to the Secretary the lands within the withdrawal under the jurisdiction of the Secretary concerned on the effective date described in subsection (j)(1).

(3) RESERVATION.—The withdrawal is reserved for use by the Secretary for development, preconstruction testing and performance confirmation, licensing, construction, management and operation, monitoring, closure, postclosure, and other activities associated with the disposal of high-level radioactive waste and spent nuclear fuel under the Nuclear Waste Policy Act of 1982 (42 U.S.C. 10101 et seq.).

(b) REVOCATION AND MODIFICATION OF PUBLIC LAND ORDERS AND RIGHTS-OF-WAY.—

(1) PUBLIC LAND ORDER REVOCATION.—Public Land Order 6802 of September 25, 1990, as extended by Public Land Order 7534, and any conditions or memoranda of understanding accompanying those land orders, are revoked.

(2) RIGHT-OF-WAY RESERVATIONS.—Project right-of-way reservations N–48602 and N–47748 of 6 January 2001, are revoked.

(c) LAND DESCRIPTION.—

(1) BOUNDARIES.—The lands and interests in lands withdrawn and reserved by this section comprise the approximately 147,000 acres of land in Nye County, Nevada, as generally depicted on the Yucca Mountain Project Map, YMP–03–024.2, entitled "Proposed Land Withdrawal" and dated July 14 21, 2005.

(2) LEGAL DESCRIPTION AND MAP.—Not later than 120 days after the date of enactment of this Act, the Secretary of the Interior shall—

(A) publish in the Federal Register a notice containing a legal description of the withdrawal; and

(B) file copies of the maps described in paragraph (1) and the legal description of the withdrawal with the Congress, the Governor of the State of Nevada, and the Archivist of the United States.

(3) TECHNICAL CORRECTIONS.—The maps and legal description referred to in this subsection have the same force and effect as if they were included in this section. The Secretary of the Interior may correct clerical and typographical errors in the maps and legal description.

(d) RELATIONSHIP TO OTHER RESERVATIONS.—The provisions of subtitle A of title XXX of the Military Lands Withdrawal Act of 1999 (sections 3011–3023 of Public Law 106–65) and of Public Land Order 2568 do not apply to the lands withdrawn and reserved for use by the Secretary under subsection (a). This Act does not apply to any other lands withdrawn for use by the Department of Defense under subtitle A of title XXX of the Military Lands Withdrawal Act of 1999.

(e) MANAGEMENT RESPONSIBILITIES.—

(1) GENERAL AUTHORITY.—The Secretary shall manage the lands withdrawn by subsection (a) consistent with the Federal Land Policy and Management Act of 1976 (43 U.S.C. 1701 et seq.), this section, and other applicable law. The Secretary shall consult with the Secretary concerned in discharging that responsibility.

(2) MANAGEMENT PLAN.—

(A) DEVELOPMENT.—The Secretary, after consulting with the Secretary concerned, shall develop a management plan for the use of the withdrawal. Within 3 years after the date of enactment of this Act, the Secretary shall submit the management plan to the Congress and the State of Nevada.

(B) PRIORITY OF YUCCA MOUNTAIN PROJECT-RELATED ISSUES.—Subject to subparagraphs (C) and (D), any use of the withdrawal for activities not associated with the Project is subject to conditions and restrictions that the Secretary considers necessary or desirable to permit the conduct of Project-related activities.

(C) DEPARTMENT OF THE AIR FORCEUSES.—The management plan may provide for the continued use by the Department of the Air Force of the portion of the withdrawal within the Nellis Air Force Base Test and Training Range under terms and conditions on which the Secretary and the Secretary of the Air Force agree concerning Air Force activities.

(D) OTHER NON-YUCCA-MOUNTAIN-PROJECT USES.—The management plan shall provide for the maintenance of wildlife habitat and shall provide that the Secretary may permit non-Project-related uses that the Secretary considers appropriate, including domestic livestock grazing and hunting and trapping in accordance with the following requirements:

(i) GRAZING.—The Secretary may permit grazing to continue where established before the effective date described in subsection (j)(1), subject to regulations, policies, and practices that the Secretary, after consulting with the Secretary of the Interior, determines to be necessary or appropriate. The management of grazing shall be conducted in accordance with applicable grazing laws and policies, including—

(I) the Act commonly known as the "Taylor Grazing Act" (43 U.S.C. 20 315 et seq.);

(II) title IV of the Federal Land Policy and Management Act of 1976 23 (43 U.S.C. 1751 et seq.); and

(III) the Public Rangelands Improvement Act of 1978 (43 U.S.C. 3 1901 et seq.).

(ii) HUNTING AND TRAPPING.—The Secretary may permit hunting and trapping within the withdrawal where established before the effective date described in subsection (k)(1), except that the Secretary, after consulting with the Secretary of the Interior and the State of Nevada, 11 may designate zones where, and establish periods when, no hunting or trapping is permitted for reasons of public safety, national security, administration, or public use and enjoyment.

(E) MINING.—

(i) IN GENERAL.—Except as provided in clause (ii), surface or subsurface mining or oil or gas production, including slant drilling from outside the boundaries of the withdrawal, is not permitted at any time on lands on or under the withdrawal. The Secretary of the Interior shall evaluate and adjudicate the validity of all unpatented mining claims on the portion of the with drawal that, on the date of enactment of this Act, was under the control of the Bureau of Land Management. The Secretary shall provide just compensation for the acquisition of any valid property right.

(ii) CIND-R-LITE MINE.—Patented Mining Claim No. 27–83–0002, covering the Cind-R-Lite Mine, shall not be affected by establishment of the withdrawal set forth in subsection (a)(1). In that event, the Secretary shall provide just compensation.

(F) LIMITED PUBLIC ACCESS.—The management plan may provide for limited public access to the portion of the withdrawal under Bureau of Land Management control on the effective date described in subsection (j)(1). Permitted uses may include continuation of the Nye County Early Warning Drilling Program, utility corridors, and other uses the Secretary, after consulting with the Secretary of the Interior, considers consistent with the purposes of the withdrawal.

(3) CLOSURE.—If the Secretary, after consulting with the Secretary concerned, determines that the health and safety of the public or the common defense and security require the closure of a road, trail, or other portion of the withdrawal, or the airspace above the withdrawal, the Secretary may effect and maintain the closure and shall provide notice of the closure.

(4) IMPLEMENTATION.—The Secretary and the Secretary concerned shall implement the management plan developed under paragraph (2) under terms and conditions on which they agree.

(f) IMMUNITY.—The United States and its departments and agencies shall be held harmless and shall not be liable for damages to persons or property suffered in the course of any mining, mineral leasing, or geothermal leasing activity conducted on the withdrawal.

(g) LAND ACQUISITION.—The Secretary may acquire lands and interests in lands within the withdrawal. Those lands and interests in lands may be acquired by donation, purchase, lease, exchange, easement, rights-of-way, or other appropriate methods using donated or appropriated funds. The Secretary of the Interior shall conduct any exchange of lands within the withdrawal for Federal lands outside the withdrawal.

(h) MATERIAL REQUIREMENTS.—Notwithstanding any other provision of law, no Federal, State, interstate, or local requirement, either substantive or procedural, that is referred to in section 6001(a) of the Solid Waste Disposal Act (42 U.S.C. 6961(a)) applies with respect to any material—

(1) as such material is transported to a repository for disposal at such repository; or

(2) as, or after, such material is disposed of in a repository.

(i) DEFINITIONS.—

(1) NUCLEAR WASTE POLICY ACT OF 1982 DEFINITIONS.—For purposes of this section, the terms "disposal", "high-level radioactive waste", "repository", "Secretary", and "spent nuclear fuel" have the meaning given those terms in section 2 of the Nuclear Waste Policy Act of 1982 (42 U.S.C. 16 10101).

(2) OTHER DEFINITIONS.—For purposes of this section—

(A) the term "withdrawal" means the geographic area consisting of the land described in subsection (c);

(B) the term "Secretary concerned" means the Secretary of the Air Force or the Secretary of the Interior, or both, as appropriate; and (C) the term "Project" means the Yucca Mountain Project.

(j) EFFECTIVE DATE.—

(1) IN GENERAL.—Except as provided in paragraph (2), this section shall take effect on the date on which the Nuclear Regulatory Commission issues a final decision approving the issuance of a construction authorization for a repository under section 114(d)(1) of the Nuclear Waste Policy Act of 1982 (42 U.S.C. 10134(d)) (as so designated by this Act).

(2) EXCEPTIONS.—Subsections (c), (e)(2)(A), (h), (i), and (j) shall take effect on the date of enactment of this Act.

Sec. 202. Application Procedures and Infrastructure Activities

(a) STATUS REPORT ON APPLICATION.—Section 114(c) of the Nuclear Waste Policy Act of 1982 (42 U.S.C. 10134(c)) is amended by striking "the date on which such authorization is granted" and inserting "the date on which the Commission issues a final decision approving or disapproving such application".

(b) APPLICATION PROCEDURES AND INFRASTRUCTURE ACTIVITIES.—Section 114(d) of the Nuclear Waste Policy Act of 1982 (42 U.S.C. 10134(d)) is amended—

(1) by striking "The Commission shall consider" and inserting the following:

"(1) APPLICATIONS FOR CONSTRUCTION AUTHORIZATION.—The Commission shall consider";

(2) by striking "the expiration of 3 years after the date of the submission of such application" and inserting "30 months after the date of enactment of the Nuclear Waste Policy Amendments Act of 9 2019";

(3) by striking "70,000 metric tons" each place it appears and inserting "110,000 metric tons"; and (4) by adding at the end the following new paragraphs:

"(2) APPLICATIONS TO AMEND.—If the Commission issues a construction authorization for a repository pursuant to paragraph (1) and the Secretary submits an application to amend such authorization, the Commission shall consider the application to amend using expedited, informal procedures, including discovery procedures that minimize the burden on the parties to produce documents. The Commission shall issue a final decision on such application to amend within 1 year after the date of submission of such application, except that the Commission may extend such deadline by not more than 6 months if, not less than 30 days before such deadline, the Commission complies with the reporting requirements established in subsection (e)(2).

"(3) INFRASTRUCTURE ACTIVITIES.—

"(A) IN GENERAL.—At any time before or after the Commission issues a final decision approving or disapproving the issuance of a construction authorization for a repository pursuant to paragraph (1), the Secretary may undertake infrastructure activities that the Secretary considers necessary or appropriate to support construction or operation of a repository at the Yucca Mountain site or transportation to such site of spent nuclear fuel and high-level radioactive waste. Infrastructure activities include safety upgrades, site preparation, the construction of a rail line to connect the Yucca Mountain site with the national rail network (including any facilities to facilitate rail operations), and construction, upgrade, acquisition, or operation of electrical grids or facilities, other utilities, communication facilities, access roads, and nonnuclear support facilities.

"(B) ENVIRONMENTAL ANALYSIS.—If the Secretary determines that an environmental analysis is required under the National Environmental Policy Act of 1969 with respect to an infrastructure activity undertaken under thisparagraph, the Secretary need not consider alternative actions or a no-action alternative. To the extent any other Federal agency must consider the potential environmental impact of such an infrastructure activity, the agency shall adopt, to the extent practicable, any environmental analysis prepared by the Secretary under this subparagraph without further action.

Such adoption satisfies the responsibilities of the adopting agency under the National Environmental Policy Act of 1969, and no further action is required by the agency.

"(C) NO GROUNDS FOR DISAPPROVAL.—

The Commission may not disapprove, on the grounds that the Secretary undertook an infrastructure activity under this paragraph—

"(i) the issuance of a construction authorization for a repository pursuant to paragraph (1);

"(ii) a license to receive and possess spent nuclear fuel and high-level radioactive waste; or

"(iii) any other action concerning the repository.".

(c) CONNECTED ACTIONS.—Section 114(f)(6) of the Nuclear Waste Policy Act of 1982 (42 U.S.C. 10134(f)(6)) is amended by striking "or nongeologic alternatives to such site" and inserting "nongeologic alternatives to such site, or an action connected or otherwise related to the repository to the extent the action is undertaken outside the geologic repository operations area and does not require a license from the Commission".

Sec. 203. Pending Repository License Application

Nothing in this Act or the amendments made by this Act shall be construed to require the Secretary to amend or otherwise modify an application for a construction authorization described in section 114(d) of the Nuclear Waste Policy Act of 1982 (42 U.S.C. 10134(d)) pending as of the date of enactment of this Act.

Sec. 204. Limitation on Planning, Development, or Construction of Defense Waste Repository

(a) LIMITATION.—The Secretary of Energy may not take any action relating to the planning, development, or construction of a defense waste repository until the date on which the Nuclear Regulatory Commission issues a final decision approving or disapproving the issuance of a construction authorization for a repository under section 114(d)(1) of the Nuclear Waste Policy Act of 1982 (42 U.S.C. 10134(d)) (as so designated by this Act).

(b) DEFINITIONS.—In this section—

(1) the terms "atomic energy defense activity", "high-level radioactive waste", "repository", and "spent nuclear fuel" have the meanings given those terms in section 2 of the Nuclear Waste Policy Act 9 of 1982 (42 U.S.C. 10101); and

(2) the term "defense waste repository" means the repository for high-level radioactive waste and spent nuclear fuel derived from the atomic energy defense activities of the Department of Energy, as described in the draft plan of the Department titled "Draft Plan for a Defense Waste Repository" published on December 16, 2016.

Sec. 205. Sense of Congress Regarding Transportation Routes

It is the sense of Congress that the Secretary of Energy should consider routes for the transportation of spent nuclear fuel or high-level radioactive waste transported by or for the Secretary under subtitle A of title I of the Nuclear Waste Policy Act of 1982 (42 U.S.C. 10131 et seq.) to the Yucca Mountain site that, to the extent practicable, avoid Las Vegas, Nevada.

Title III—DOE Contract Performance

Sec. 301. Title to Material.

Section 123 of the Nuclear Waste Policy Act of 1982 5 (42 U.S.C. 10143) is amended—

(1) by striking "Delivery" and inserting "(a) IN GENERAL.—Delivery";

(2) by striking "repository constructed under this subtitle" and inserting "repository or monitored retrievable storage facility"; and

(3) by adding at the end the following new sub-section:

"(b) CONTRACT MODIFICATION.—The Secretary may enter into new contracts or negotiate modifications to existing contracts, with any person who generates or holds title to high-level radioactive waste or spent nuclear fuel of domestic origin, for acceptance of title, subsequent transportation, and storage of such high-level radioactive waste or spent nuclear fuel (including to expedite such acceptance of title, transportation, and storage of such wasteor fuel from facilities that have ceased commercial operation) at a monitored retrievable storage facility authorized under subtitle C.".

Title IV—Benefits to Host Community

Sec. 401. Consent.

Section 170 of the Nuclear Waste Policy Act of 1982 5 (42 U.S.C. 10173) is amended—

(1) in subsection (c), by striking "shall offer" and inserting "may offer";

(2) in subsection (d), by striking "shall" and inserting "may";

(3) in subsection (e)—

(A) by inserting a comma after "repository"; and

(B) by inserting "per State," after "facility"; and

(4) by adding at the end the following new subsection:

"(g) CONSENT.—The acceptance or use of any of the benefits provided under a benefits agreement under this section by the State of Nevada shall not be considered to be an expression of consent, express or implied, to the siting of a repository in such State.".

Sec. 402. Content of Agreements.

(a) BENEFITS SCHEDULE.—The table in section 171(a)(1) of the Nuclear Waste Policy Act of 1982 (42 U.S.C. 10173a(a)(1)) is amended to read as follows:

Benefits Schedule

Event	MRS	Repository
(A) Annual payments prior to first spent fuel receipt	$5,000,000	$15,000,000
(B) Upon first spent fuel receipt	$10,000,000	$400,000,000
(C) Annual payments after first spent fuel receipt until closure of the facility	$10,000,000	$40,000,000".

(b) RESTRICTIONS ON USE.—Section 171(a) of the Nuclear Waste Policy Act of 1982 (42 U.S.C. 10173a(a)) is amended—

(1) in paragraph (6), by striking "paragraph (7)" and inserting "paragraphs (7) and (8)"; and (2) by adding at the end the following new paragraph:

"(8) None of the payments under this section may be used—

"(A) directly or indirectly to influence legislative action on any matter pending before Congress or a State legislature or for any lobbying activity as provided in section 1913 of title 18, United States Code;

"(B) for litigation purposes; or

"(C) to support multistate efforts or other coalition-building activities inconsistent with the siting, construction, or operation of the monitored retrievable storage facility or repository concerned.".

(c) CONTENTS.—Section 171(b) of the Nuclear Waste Policy Act of 1982 (42 U.S.C. 10173a(b)) is amended—

(1) by striking paragraph (2);

(2) by redesignating paragraphs (3) through (5) as paragraphs (2) through (4), respectively; and

(3) in paragraph (3) (as redesignated by paragraph (2) of this subsection), by striking "in the design of the repository or monitored retrievable storage facility and".

(d) PAYMENTS BY SECRETARY.—Section 171(c) of the Nuclear Waste Policy Act of 1982 (42 U.S.C. 10173a(c)) is amended to read as follows:

"(c) PAYMENTS BY SECRETARY.—The Secretary shall make payments to the State of Nevada under a benefits agreement concerning a repository under section 170 from the Waste Fund. The signature of the Secretary on a valid benefits agreement under this subtitle shall constitute a commitment, but only to the extent that all amounts for that purpose are provided in advance in subsequent appropriations Acts, by the Secretary to make payments in accordance with such agreement.".

Sec. 403. Covered Units of Local Government.

(a) IN GENERAL.—The Nuclear Waste Policy Act of 1982 (42 U.S.C. 10101 et seq.) is amended by inserting after section 172 the following new section:

"Sec. 172A. Covered Units of Local Government.

"(a) BENEFITS AGREEMENT.—Not earlier than 1 year after the date of enactment of this section, the Secretary may enter into a benefits agreement with any covered unit of local government concerning a repository for the acceptance of high-level radioactive waste or spent nuclear fuel in the State of Nevada.

"(b) CONTENT OF AGREEMENTS.—In addition to any benefits that a covered unit of local government may receive under this Act, the Secretary shall make payments to such covered unit of local government that is a party to a benefits agreement under subsection (a) to mitigate impacts described in section 175(b).

"(c) PAYMENTS FROM WASTE FUND.—The Secretary shall make payments to a covered unit of local government under a benefits agreement under this section from the Waste Fund.

"(d) RESTRICTION ON USE.—None of the payments made pursuant to a benefits agreement under this section 24 may be used—

"(1) directly or indirectly to influence legislative action on any matter pending before Congress or a State legislature or for any lobbying activity as provided in section 1913 of title 18, United States Code;

"(2) for litigation purposes; or

"(3) to support multistate efforts or other coalition-building activities inconsistent with the siting, construction, or operation of the repository.

"(e) CONSENT.—The acceptance or use of any of the benefits provided under a benefits agreement under this section by any covered unit of local government shall not be considered to be an expression of consent, express or implied, to the siting of a repository in the State of Nevada.

"(f) COVERED UNIT OF LOCAL GOVERNMENT DEFINED.—In this section, the term 'covered unit of local government' means—

"(1) any affected unit of local government with respect to a repository; and

"(2) any unit of general local government in the State of Nevada.".

(b) CONFORMING AMENDMENTS.—

(1) BENEFITS AGREEMENT.—Section 170(a)(4) of the Nuclear Waste Policy Act of 1982 (42 U.S.C. 10173(a)(4)) is amended to read as follows:

"(4) Benefits and payments under this subtitle made available pursuant to a benefits agreement under this section or section 172A may be made available only in accordance with such benefits agreement and to the extent that all amounts for that purpose are provided in advance in subsequent appropriations Acts.".

(2) LIMITATION.—Section 170(e) of the Nuclear Waste Policy Act of 1982 (42 U.S.C. 10173(e)) is further amended by inserting "under this section" after "may be in effect".

(3) TABLE OF CONTENTS.—The table of contents for the Nuclear Waste Policy Act of 1982 (42 U.S.C. 10101 note) is amended by adding after the item relating to section 172, the following: "Sec. 172A. Covered units of local government.".

Sec. 404. Termination.

Section 173 of the Nuclear Waste Policy Act of 1982 (42 U.S.C. 10173c) is amended—

(1) in subsection (a)—

(A) by striking "under this title if" and inserting "under this title";

(B) in paragraph (1), by inserting "concerning a repository or a monitored retrievable storage facility, if" before "the site under consideration"; and

(C) in paragraph (2), by striking "the Secretary determines that the Commission cannot license the facility within a reasonable time" and inserting "concerning a repository, if the Commission issues a final decision disapproving the issuance of a construction authorization fora repository under section 114(d)(1)"; and (2) by amending subsection (b) to read as follows:

"(b) TERMINATION BY STATE OR INDIAN TRIBE.—

A State, covered unit of local government (as defined insection 172A), or Indian tribe may only terminate a benefits agreement under this title—

"(1) concerning a repository or a monitored retrievable storage facility, if the Secretary disqualifies the site under consideration for its failure to comply with technical requirements established by the Secretary in accordance with this Act; or

"(2) concerning a repository, if the Commission issues a final decision disapproving the issuance of a construction authorization for a repository under 22 section 114(d)(1).".

Sec. 405. Priority Funding for Certain Institutions of Higher Education.

(a) IN GENERAL.—Subtitle G of the Nuclear Waste Policy Act of 1982 (42 U.S.C. 10174 et seq.) is amended by adding at the end the following new section:

"Sec. 176. Priority Funding for Certain Institutions of Higher Education.

"(a) IN GENERAL.—In providing any funding to institutions of higher education from the Waste Fund, the Secretary shall prioritize institutions of higher education that are located in the State of Nevada.

"(b) DEFINITION.—In this section, the term 'institution of higher education' has the meaning given that term in section 101 of the Higher Education Act of 1965 (20 15 U.S.C. 1001).".

(b) CONFORMING AMENDMENT.—The table of contents for the Nuclear Waste Policy Act of 1982 (42 U.S.C. 10101 note) is amended by adding after the item relating to section 175, the following:

"Sec. 176. Priority funding for certain institutions of higher education.".

Sec. 406. Disposal of Spent Nuclear Fuel.

Section 122 of the Nuclear Waste Policy Act of 1982 (42 U.S.C. 10142) is amended by adding at the end the following: "Any economic benefits derived from the retrieval of spent nuclear fuel pursuant to this section shall be shared with the State in which the repository is located, affected units of local government, and affected Indian tribes.".

Sec. 407. Updated Report.

Section 175(a) of the Nuclear Waste Policy Act of 1982 (42 U.S.C. 10174a(a)) is amended by striking "Nuclear Waste Policy Amendments Act of 1987" and inserting "Nuclear Waste Policy Amendments Act of 2019".

Title V—Funding

Sec. 501. Assessment and Collection of Fees.

(a) IN GENERAL.—Section 302(a)(4) of the NuclearWaste Policy Act of 1982 (42 U.S.C. 10222(a)(4)) isamended—

(1) in the first sentence—

(A) by striking "(4) Not later than" and inserting the following:

"(4) ASSESSMENT, COLLECTION, AND PAYMENT OF FEES.—

"(A) ASSESSMENT OF FEES.—Not later than";

(B) by striking "the date of enactment of this Act" and inserting "the date of enactment of the Nuclear Waste Policy Amendments Act 23 of 2019"; and

(C) by striking "collection and payment" and inserting "assessment";

(2) in the second sentence, by striking "collection of the fee" and inserting "such amount";

(3) in the third sentence, by striking "are being collected" and inserting "will result from such amounts";

(4) in the fifth sentence, by striking "a period of 90 days of continuous session" and all that follows through the period at the end and inserting "the date that is 180 days after the date of such transmittal."; and

(5) by adding at the end the following:

"(B) COLLECTION AND PAYMENT OF FEES.—

"(i) IN GENERAL.—Not later than 180 days after the date of enactment of Nuclear Waste Policy Amendments Act of 2019, the Secretary shall establish procedures for the collection and payment of the fees established by paragraph (2) and paragraph (3), or adjusted pursuant to subparagraph (A).

"(ii) LIMITATION ON COLLECTION.—

The Secretary may not collect a fee established under paragraph (2), including a fee established under paragraph (2) and adjusted pursuant to subparagraph (A)—

"(I) until the date on which the Commission issues a final decision approving or disapproving the issuance of a construction authorization for a repository under section 114(d)(1);

and

"(II) after such date, in an amount that will cause the total amount of fees collected under this subsection in any fiscal year to exceed 90 percent of the amounts appropriated for that fiscal year for purposes described in subsection (d).

The limitation in subclause (II) shall not apply during a fiscal year if, at any time during that fiscal year, the Waste Fund has a balance of zero.

"(iii) PAYMENT OF FULL AMOUNTS.—

Notwithstanding the noncollection of a fee by the Secretary pursuant to clause (ii) in any fiscal year, a person who has entered into a contract with the Secretary under this subsection shall pay any uncollected amounts when determined necessary by the Secretary, subject to clause (ii), for purposes described in subsection (d).".

(b) AUTHORITY TO MODIFY CONTRACTS.—The Secretary of Energy may seek to modify a contract entered into under section 302(a) of the Nuclear Waste Policy Act of 1982 (42 U.S.C. 10222(a)) before the date of enactment of this Act to ensure that the contract complies with the provisions of such section, as amended by this Act.

(c) TECHNICAL AND CONFORMING AMENDMENTS.—

Section 302(a) of the Nuclear Waste Policy Act of 1982 12 (42 U.S.C. 10222(a)) is amended—

(1) in paragraph (1), by striking "paragraphs

(2) and (3)" and inserting "paragraphs (2), (3), and 15 (4)";

(2) in paragraph (3), by striking "126(b)"; and

(3) in paragraph (4), by striking "insure" and inserting "ensure".

Sec. 502. Use of Waste Fund.

(a) IN GENERAL.—Section 302(d) of the Nuclear Waste Policy Act of 1982 (42 U.S.C. 10222(d)) is amended—

(1) in paragraph (1), by striking "maintenance and monitoring" and all that follows through the semicolon at the end and inserting "maintenance and monitoring of any repository or test and evaluation facility constructed under this Act;";

(2) in paragraph (4), by striking "to be disposed of" and all that follows through the semicolon at the end and inserting "to be disposed of in a repository or to be used in a test and evaluation facility;";

(3) in paragraph (5), by striking "at a repository site" and all that follows through the end and inserting "at a repository site or a test and evaluation facility site and necessary or incident to such repository or test and evaluation facility;";

(4) in paragraph (6), by striking the period at the end and inserting "; and"; and

(5) by inserting after paragraph (6) the following:

"(7) payments under benefits agreements for a repository entered into under section 170 or 172A.".

(b) CONFORMING AMENDMENTS.—Section 117(d) of the Nuclear Waste Policy Act of 1982 (42 U.S.C. 10137(d)) is amended by inserting "designated with respect to a repository" after "such representatives".

Sec. 503. Annual Multiyear Budget Proposal.

Section 302(e)(2) of the Nuclear Waste Policy Act of 1982 (42 U.S.C. 10222(e)(2)) is amended by striking "triennially" and inserting "annually".

Sec. 504. Availability of Certain Amounts.

Section 302 of the Nuclear Waste Policy Act of 1982 (42 U.S.C. 10222) is amended by adding at the end the following:

"(f) LIMITATION ON FUNDING.—

"(1) IN GENERAL.—Beginning on the date of first spent fuel receipt at a repository, no amount 12 may be appropriated in any fiscal year for activities relating to the repository, including transportation of additional spent fuel to the repository and operation of the repository, unless the applicable amount required with respect to the repository under section 171(a)(1)(B) or section 171(a)(1)(C) is appropriated for that fiscal year.

"(2) DEFINITION.—In this subsection, the terms 'spent fuel' and 'first spent fuel receipt' have the meaning given such terms in section 171(a).

"(g) OFFSETTING FUNDING.—

"(1) IN GENERAL.—Fees collected after the date of enactment of the Nuclear Waste Policy Amendments Act of 2019 pursuant to subsection (a) shall be credited to the Waste Fund and available, to the extent provided in advance in appropriation Acts and consistent with the requirements of this section, to carry out activities authorized to be funded from the Waste Fund.

"(2) OFFSETTING COLLECTION.—Fees collected in a fiscal year pursuant to paragraph (1) shall be deposited and credited as offsetting collections to the account providing appropriations for such activities and shall be classified as discretionary appropriations as defined by section 250(c)(7) of the Balanced Budget and Emergency Deficit Control Act of 1985 12 (2 U.S.C. 900(c)(7)).

"(3) ESTIMATES.—For the purposes of the Balanced Budget and Emergency Deficit Control Act of 1985 (2 U.S.C. 900 et seq.) and the Congressional Budget Act of 1974 (2 U.S.C. 621 et seq.) and for determining points of order pursuant to that Act or any concurrent resolution on the budget, an estimate provided under those Acts for a provision in a bill or joint resolution, or amendment thereto or conference report thereon, that provides discretionary appropriations, derived from amounts in the Waste Fund, for such activities shall include in that estimate the amount of such fees that will be collected during the fiscal year for which such appropriation is made available. Any such estimate shall not include any change in net direct spending as result in the appropriation of such fees.".

Title VI—Miscellaneous

Sec. 601. Certain Standards and Criteria.

(a) GENERALLY APPLICABLE STANDARDS AND CRITERIA.—

(1) ENVIRONMENTAL PROTECTION AGENCY STANDARDS.—

(A) DETERMINATION AND REPORT.—Not later than 2 years after the Nuclear Regulatory Commission has issued a final decision approving or disapproving the issuance of a construction authorization for a repository under section 114(d)(1) of the Nuclear Waste Policy Act of 16 1982 (42 U.S.C. 10134(d)) (as so designated by this Act), the Administrator of the Environmental Protection Agency shall—

(i) determine if the generally applicable standards promulgated under section 121(a) of the Nuclear Waste Policy Act of 22 1982 (42 U.S.C. 10141(a)) should be updated; and

(ii) submit to Congress a report on such determination.

(B) RULE.—If the Administrator of the Environmental Protection Agency determines, under subparagraph (A), that the generally applicable standards promulgated under section 121(a) of the Nuclear Waste Policy Act of 1982 (42 U.S.C. 10141(a)) should be updated, the Administrator, not later than 2 years after submission of the report under subparagraph (A)(ii), shall, by rule, promulgate updated generally applicable standards under such section.

(2) COMMISSION REQUIREMENTS AND CRITERIA.—Not later than 2 years after the Administrator of the Environmental Protection Agency promulgates updated generally applicable standards pursuant to paragraph (1)(B), the Commission shall, by rule, promulgate updated technical requirements and criteria under section 121(b) of the Nuclear Waste Policy Act of 1982 (42 U.S.C. 10141(b)) as necessary to be consistent with such updated generally applicable standards.

(b) SITE-SPECIFIC STANDARDS AND CRITERIA.—

Nothing in this section shall affect the standards, technical requirements, and criteria promulgated by the Administrator of the Environmental Protection Agency and the Nuclear Regulatory Commission for the Yucca Mountain site under section 801 of the Energy Policy Act of 1992 (42 U.S.C. 10141 note).

Sec. 602. Application.

Section 135 of the Nuclear Waste Policy Act of 1982 (42 U.S.C. 10155) is amended by striking subsection (h) and redesignating subsection (i) as subsection (h).

Sec. 603. Transportation Safety Assistance.

Section 180(c) of the Nuclear Waste Policy Act of 9 1982 (42 U.S.C. 10175(c)) is amended—

(1) by striking "(c) The Secretary" and inserting the following:

"(c) TRAINING AND ASSISTANCE.—

"(1) TRAINING.—The Secretary"; and (2) by striking "The Waste Fund" and inserting the following:

"(2) ASSISTANCE.—The Secretary shall, subject to the availability of appropriations, provide in-kind, financial, technical, and other appropriate assistance, for safety activities related to the transportation of high-level radioactive waste or spent nuclear fuel, to any entity receiving technical assistance or funds under paragraph (1).

"(3) SOURCE OF FUNDING.—The Waste Fund".

Sec. 604. Office of Spent Nuclear Fuel.

(a) AMENDMENT TO THE NUCLEAR WASTE POLICY ACT OF 1982.—Section 304 of the Nuclear Waste Policy Act of 1982 (42 U.S.C. 10224(b)) is amended—

(1) in the heading, by striking "OFFICE OF CIVILIAN RADIOACTIVE WASTE MANAGEMENT" and inserting "OFFICE OF SPENT NUCLEAR FUEL";

(2) in subsection (a), by striking "Office of Civilian Radioactive Waste Management" and inserting "Office of Spent Nuclear Fuel"; and

(3) by amending subsection (b) to read as follows:

"(b) DIRECTOR.—

"(1) FUNCTIONS.—The Director of the Office shall be responsible for carrying out the functions of the Secretary under this Act. The Director of the Office shall report directly to the Secretary.

"(2) QUALIFICATIONS.—The Director of the Office shall be appointed from among persons who have extensive expertise and experience in organizational and project management.

"(3) REMOVAL.—The President may remove the Director only for inefficiency, neglect of duty, or malfeasance in office. If the President removes the Director, the President shall submit to Congress a statement explaining the reason for such removal.".

(b) TRANSFER OF FUNCTIONS.—

(1) AMENDMENT.—Section 203(a) of the Department of Energy Organization Act (42 U.S.C. 7133(a)) is amended by striking paragraph (8).

(2) TRANSFER OF FUNCTIONS.—The functions described in the paragraph (8) stricken by the amendment made by paragraph (1) shall be transferred to and performed by the Office of Spent Nuclear Fuel, as provided in section 304 of the Nuclear Waste Policy Act of 1982 (42 U.S.C. 10224).

(c) TECHNICAL AND CONFORMING AMENDMENTS.—

The Nuclear Waste Policy Act of 1982 is amended—

(1) in the table of contents, by amending the item relating to section 304 to read as follows:

"Sec. 304. Office of Spent Nuclear Fuel.";

and

(2) in section 2(17) (42 U.S.C. 10101(17)), by striking "Office of Civilian Radioactive Waste Management established in section 305" and inserting "Office of Spent Nuclear Fuel established in section 20 304".

(d) REFERENCES.—Any reference to the Office of Civilian Radioactive Waste Management in any law, regulation, document, record, executive order, or other paper of the United States shall be deemed to be a reference to the Office of Spent Nuclear Fuel.

Sec. 605. Subseabed or Ocean Water Disposal.

(a) PROHIBITION.—Section 5 of the Nuclear Waste Policy Act of 1982 (42 U.S.C. 10104) is amended—

(1) by striking "Nothing in this Act" and inserting:

"(a) EFFECT ON MARINE PROTECTION, RESEARCH, AND SANCTUARIES ACT OF 1972.—Nothing in this Act";

and

(2) by adding at the end the following new subsection:

"(b) SUBSEABED OR OCEAN WATER DISPOSAL.— Notwithstanding any other provision of law—

"(1) the subseabed or ocean water disposal of spent nuclear fuel or high-level radioactive waste is prohibited; and

"(2) no funds shall be obligated for any activity relating to the subseabed or ocean water disposal of spent nuclear fuel or high-level radioactive waste.".

(b) REPEAL.—Section 224 of the Nuclear Waste Policy Act of 1982, and the item relating thereto in the table of contents for such Act, are repealed.

Sec. 606. Budgetary Effects.

(a) STATUTORY PAYGO SCORECARDS.—The budgetary effects of this Act and the amendments made by this Act shall not be entered on either PAYGO scorecard maintained pursuant to section 4(d) of the Statutory Pay-As-You-Go Act of 2010.

(b) SENATE PAYGO SCORECARDS.—The budgetary effects of this Act and the amendments made by this Act shall not be entered on any PAYGO scorecard maintained for purposes of section 4106 of H. Con. Res. 71 (115th Congress).

Sec. 607. Requirement for Financial Statements Summary.

The Department of Energy shall include a financial statements summary in each audit report on the Department of Energy Nuclear Waste Fund's fiscal year financial statement audit.

Sec. 608. Stranded Nuclear Waste.

(a) STRANDED NUCLEAR WASTE TASK FORCE.—

(1) ESTABLISHMENT.—The Secretary shall establish a task force, to be known as the Stranded Nuclear Waste Task Force—

(A) to conduct a study on existing public and private resources and funding for which affected communities may be eligible; and

(B) to develop immediate and long-term economic adjustment plans tailored to the needs of each affected community.

(2) STUDY.—Not later than 180 days after the date of enactment of this Act, the Stranded Nuclear Waste Task Force shall complete and submit to Congress the study described in paragraph (1).

(b) DEFINITIONS.—In this section:

(1) AFFECTED COMMUNITY.—The term "affected community" means a municipality that contains stranded nuclear waste within the boundaries of the municipality, as determined by the Secretary.

(2) SECRETARY.—The term "Secretary" means the Secretary of Energy.

(3) STRANDED NUCLEAR WASTE.—The term "stranded nuclear waste" means nuclear waste or spent nuclear fuel stored in dry casks or spent fuel pools at a decommissioned or decommissioning nuclear facility.

116th Congress, 1st Session, H. R. L29195

To amend the Nuclear Waste Policy Act of 1982 to prioritize the acceptance of high-level radioactive waste or spent nuclear fuel from certain civilian nuclear power reactors, and for other purposes.

May 23, 2019

IN THE HOUSE OF REPRESENTATIVES

Mr. LEVIN of California introduced the following bill; which was referred to the Committee on Enlergy and Commerce.

A Bill

To amend the Nuclear Waste Policy Act of 1982 to prioritize the acceptance of high-level radioactive waste or spent nuclear fuel from certain civilian nuclear power reactors, and for other purposes.

Be it enacted by the Senate and House of Representatives of the United States of America in Congress assembled,

Section 1. Short Title.

This Act may be cited as the "Spent Fuel Prioritization Act of 2019".

Sec. 2. Acceptance Priority for High-Level Radioactive Waste and Spent Nuclear Fuel.

Section 302(a) of the Nuclear Waste Policy Act of 1982 (42 U.S.C. 10222(a)) is amended by adding at the end the following:

"(7) In determining the order in which the Secretary will accept high-level radioactive waste or spent nuclear fuel for disposal or storage, the Secretary shall prioritize waste or spent fuel from a civilian nuclear power reactor based on the operating status of the reactor, the population of the area in which the reactor is located, and the earthquake hazard of the area in which the reactor is located, giving highest priority to a reactor that is—

"(A) decomissioned or decommissioning;

"(B) located in an area, as determined by the Secretary, the population of which is the largest; and

"(C) located in an area with the highest hazard of an earthquake, as indicated by the Seismic Hazard Maps published by the Director of the United States Geological Survey pursuant to section 5(b)(3)(J) of the Earthquake Hazards Reduction 22 Act of 1977.".

116th Congress, 1st Session, H. R. 3136

To direct the Secretary of Energy to establish a program for the interim storage of high-level radioactive waste and spent nuclear fuel, and for other purposes.

June 5, 2019

IN THE HOUSE OF REPRESENTATIVES

Ms. MATSUI introduced the following bill; which was referred to the Committee on Energy and Commerce.

A Bill

To direct the Secretary of Energy to establish a program for the interim storage of high-level radioactive waste and spent nuclear fuel, and for other purposes.

Be it enacted by the Senate and House of Representatives of the United States of America in Congress assembled,

Section 1. Short Title.

This Act may bc citcd as thc "Storagc and Transportation of Residual and Excess Nuclear Fuel Act of 2019", or the "STORE Nuclear Fuel Act of 2019".

Sec. 2. Interim Storage.

(a) IN GENERAL.—Title I of the Nuclear Waste Policy Act of 1982 (42 U.S.C. 10121 et seq.) is amended by adding at the end the following:

"Subtitle I—Interim Storage

"Sec. 190. Definitions.

"In this subtitle:

"(1) CONTRACT HOLDER.—The term 'contract holder' means any person who—

"(A) generates or holds title to spent nuclear fuel and high-level radioactive waste generated at a civilian nuclear power reactor; and

"(B) has entered into a contract for the disposal of spent nuclear fuel and high-level radioactive waste under section 302(a).

"(2) EMERGENCY DELIVERY.—

"(A) IN GENERAL.—The term 'emergency delivery' means spent nuclear fuel and high-level radioactive waste accepted by the Secretary for storage prior to the date provided in the contractual delivery commitment schedule of the standard contract for disposal of spent nuclear fuel and radioactive waste pursuant to section 302(a).

"(B) INCLUSION.—The term 'emergency delivery' may include, at the discretion of the Secretary, spent nuclear fuel and high-level radioactive waste generated by an atomic energy defense activity that is required to be removed from a Department of Energy facility—

"(i) pursuant to a compliance agreement; or

"(ii) to eliminate an imminent and serious threat to the health and safety of the public or the common defense and security.

"(3) PRIORITY WASTE.—The term 'priority waste' means—

"(A) any emergency delivery; and

"(B) spent nuclear fuel or high-level radioactive waste from a civilian nuclear power reactor that has been permanently shut down.

"(4) STORAGE FACILITY.—The term 'storage facility' means a facility for the consolidated storage of spent nuclear fuel and high-level radioactive waste from multiple contract holders or the Secretary pending the disposal of the spent nuclear fuel and high-level radioactive waste in a repository.

"Sec. 191. Program for Storage Facilities.

"(a) ESTABLISHMENT OF PROGRAM.—The Secretary shall establish a program under which the Secretary may—

"(1) site, construct, and operate one or more storage facilities licensed by the Commission under the Atomic Energy Act of 1954; and

"(2) store, pursuant to a storage contract, high-level radioactive waste or spent nuclear fuel at a storage facility for which a non-Federal entity holds a license issued by the Commission under such Act.

"(b) INTERIM STORAGE AGREEMENTS AUTHORIZED.—

"(1) IN GENERAL.—The Secretary may enter into an agreement with any contract holder for acceptance of title pursuant to section 302(a), subse-

quent transportation, and interim storage of high-level radioactive waste or spent nuclear fuel (including to expedite such acceptance of title, transportation, and storage of such waste or spent fuel from facilities that have ceased commercial operation) at a storage facility under this section.

"(2) PRIORITY WASTE.—In entering into agreements under paragraph (1), the Secretary shall prioritize acceptance of priority waste.

"(c) PRIORITY.—

"(1) IN GENERAL.—Except as provided in paragraph (2), the Secretary shall prioritize storage authorized under subsection (a)(2).

"(2) EXCEPTION.—

"(A) DETERMINATION.—Paragraph (1) shall not apply if the Secretary determines that it will be faster and less expensive to site, construct, and operate a facility authorized under subsection (a)(1), in comparison with a facility authorized under subsection (a)(2).

"(B) NOTIFICATION.—Not later than 30 days after the Secretary makes a determination described in subparagraph (A), the Secretary shall submit to Congress written notification of such determination.

"(d) REQUEST FOR PROPOSALS.—

"(1) IN GENERAL.—Not later than 180 days after the date of enactment of this subtitle, the Secretary shall issue a request for proposals for storage authorized under subsection (a)(2)—

"(A) to obtain any license from the Commission and any other Federal or State entity that is necessary for the construction of 1 or more storage facilities;

"(B) to safely transport spent nuclear fuel and high-level radioactive waste, as applicable, to such storage facilities; and

"(C) to safely store spent nuclear fuel and high-level radioactive waste, as applicable, at such storage facilities, pending the construction and operation of a repository.

"(2) GUIDELINES.—

"(A) IN GENERAL.—The request for proposals under paragraph (1) shall include general guidelines for storage facilities consistent with each requirement of section 112(a) that the Secretary determines to be applicable to storage under this section.

"(B) REVISIONS.—The Secretary may revise the general guidelines as necessary, consistent with this section.

"(e) REVIEW OF PROPOSALS.—The Secretary shall review each proposal submitted pursuant to subsection (d) to evaluate—

"(1) the extent to which the applicable States, affected units of local government, and affected Indian tribes support the proposal;

"(2) the likelihood that the proposed site for the storage facility is suitable for site evaluation under the guidelines included under subsection 24 (d)(2);

"(3) a reasonable comparative evaluation of the proposed site and other proposed sites;

"(4) the extent to which spent nuclear fuel and high-level radioactive waste are, or are planned to be, stored or disposed of within the State;

"(5) the extent to which the proposal would—

"(A) enhance the reliability and flexibility of the system for the disposal of spent nuclear fuel and high-level radioactive waste, including co-location with a proposed repository; and

"(B) minimize the effects on the public of transportation and handling of spent nuclear fuel and high-level radioactive waste;

"(6) potential conflicts with—

"(A) any compliance agreement requiring removal of spent nuclear fuel and high-level radioactive waste from a site; or

"(B) a statutory prohibition on the storage or disposal of spent nuclear fuel and high-level radioactive waste at a site; and

"(7) any other criteria, including criteria relating to technical or safety specifications, that the Secretary determines to be appropriate.

"(f) SITE SELECTION.—

"(1) DETERMINATION OF SUITABILITY.—After conducting a review under subsection (e) and any additional site investigation that the Secretary determines to be appropriate, the Secretary shall determine whether a site is suitable for site evaluation under the guidelines included under subsection 7 (d)(2).

"(2) SELECTION OF SITE FOR EVALUATION.—

From the sites determined to be suitable for site evaluation under paragraph (1), the Secretary shall select at least 1 site for site evaluation, giving priority to sites that have been proposed to be co-located with a repository, after—

"(A) holding a public hearing in the vicinity of each site; and

"(B) notifying Congress.

"(3) COOPERATIVE AGREEMENT.—On selection of a site for evaluation under paragraph (2), the Secretary may enter into a cooperative agreement with the State, affected units of local government, and affected Indian tribes, as applicable, that includes—

"(A) terms of financial and technical assistance to enable each applicable unit of government to monitor, review, evaluate, comment on, obtain information on, make recommendations on, and mitigate any effects from, site evaluation activities; and

"(B) any other term that the Secretary determines to be appropriate.

"(4) CONSENT-BASED APPROVAL.—

"(A) IN GENERAL.—If the Secretary determines, based on site evaluation under this subsection, that a site is suitable for developing a storage facility, the Secretary may select the site for developing such a facility if the Secretary enters into a consent agreement with—

"(i) the State in which the site is proposed to be located;

"(ii) each affected unit of local government; and

"(iii) any affected Indian tribe.

"(B) BINDING EFFECT.—A consent agreement entered into under subparagraph (A)—

"(i) shall be binding on the parties;

and

"(ii) shall not be amended or revoked except by mutual agreement of the parties.".

(b) CONFORMING AMENDMENT.—The table of contents for the Nuclear Waste Policy Act of 1982 (42 U.S.C. 10101 note) is amended by adding after the item relating to section 180 the following:

"SUBTITLE I—INTERIM STORAGE

"Sec. 190. Definitions.

"Sec. 191. Program for storage facilities.".

Sec. 3. Limitation on Collection of Fees.

Section 302(a)(4) of the Nuclear Waste Policy Act 7 of 1982 (42 U.S.C. 10222(a)(4)) is amended—

(1) in the first sentence, by striking "(4) Not later than" and inserting the following:

"(4) COLLECTION AND PAYMENT OF FEES.—

"(A) IN GENERAL.—Not later than"; and

(2) by adding at the end the following:

"(B) LIMITATION ON COLLECTION.—The Secretary may not collect a fee established under paragraph (2), including a fee established under paragraph (2) and adjusted pursuant to subparagraph (A), until the date on which the Commission issues a final decision approving or disapproving the issuance of a construction authorization for a repository under section 21 114(d).".

Sec. 4. Funding.

Section 302(d) of the Nuclear Waste Policy Act of 3 1982 (42 U.S.C. 10222(d)) is amended—

(1) in paragraph (5), by striking "; and" and inserting a semicolon;

(2) in paragraph (6), by striking the period at the end and inserting a semicolon; and

(3) by inserting after paragraph (6) the following:

"(7) carrying out subtitle I of title I, other than consent agreements under section 191(f)(4), except that the Secretary may not expend for such purpose in a fiscal year amounts totaling more than 25 percent of the interest generated by the Fund in such fiscal year; and

"(8) consent agreements under section 17 191(f)(4).".

Edward F. Sproat III 2600, Monocacy Ford Road, Frederick, MD. 21701

June 11, 2019
Hon. Paul Tonko, Chairman
Hon. John Shimkus, Ranking Member
U.S. House Subcommittee on Environment and Climate Change
2125 Rayburn House Office Building
Washington, DC. 20515

Subject: June 13 Hearing on "Cleaning Up Communities: Options for the Storage and Disposal of Spent Nuclear Fuel"

Dear Chairman Tonko and Ranking Member Shimkus:

I'm writing to you as the last Senat e-confirmed Director of the Office of Civilian Radioactive Waste Management in the Department of Energy to thank you both for holding the subject hearing on the several proposed bills that are attempting to address the issue of stranded spent nuclear fuel and high-level nuclear waste. I held that position from June 2006 until January 2009, during which I led the effort to complete and submit the license application for the Yucca Mountain repository as required by the Nuclear Waste Policy Act (NWPA), as amended. I signed the application to the U.S. Nuclear Regulatory Commission (NRC) for the US DOE in June, 2008.

I strongly support the House review and debate of these import ant legislative initiatives. I was a witness at the House hearing on H.R. 3053 in April 2017 and have attached a copy of my testimony at that hearing for your information. It contains the rationale as to why a number of the legislation's provisions are needed to be able to move forward with 1) the final adjudication of the Yucca Mountain license application and 2) examin e alternatives should the construction license for the repository not be granted.

The possibility of the NRC denying a construction license to DOE for the Yucca site has always been a possibility, but the NWPA is silent on that potential outcome. For that reason, I am in favor of allowin g the DOE to pursue options for interim storage. If the House deems it appropriate to authorize an interim storage option, I suggest that it be conditioned on using the "consent -based" siting criteria advocated by Mr. Fettus of the Natural Resources Defense Council (NRDC) and by the State of Nevada as represented by Mr. Halstead.

I need to be clear: I do not believe consent-basedsiting will ever result in spent nuclear fuel being permanently disposed of or moved. It has been tried severaltimes. Witness the failures of the Nuclear Waste Negotiator authorized in the NWPA, the Private Spent Fuel Storage project in Utah, and the support of host Nye County, Nevada for the Yucca Repository. In all of these cases, political opposition developed around t hose sites, both in the potential host st ate as well as neighboring states through which the spent fuel would be transported. It is much easier politically to be opposed to nuclear waste shipments than to be for them. And because of the long time frames needed to evaluate then license and build a repository or interim storage site, political opposition will surely result in changes in political and elected leadership that will withdraw any consent that might be needed. But in addition to consent, the final site must also be geologically sound, which means it must be evaluated in depth before it is deemed suitable. Consent will be needed to evaluate a site and then again to proceed with licensing if it is found suitable. And still, after all of that, anyone can intervene in the licensing process regardless of who has consented. Consent-based siting does sound very appealing. I just don't see it leading to a successful conclusion. Of course, I may be wrong. So why not let the proposed interim storage site(s) try the path proposed by the NRDC and Nevada in their testimony and let's see if it will work.

I would be more than happy to meet or discuss with you or your staffs any questions you may have regarding this letter and my positions in it.

If you both deem fit to allow it, I respectfully request that this letter and its attachment be submitted to the record of the subject hearing.

Very Respectfully,

Edward F. Sproat III
Former Director
OCRWM, US DOE

Statement of Edward F. Sproat III, Former Director (RW-1), Office of Civilian Radioactive Waste Management, U.S. Department of Energy Before the Subcommittee on Environment, Committee on Energy and Commerce, U.S. House of Representatives, April 26, 2017

Mr. Chairman and Members of the Committee, I am very pleased to have been requested to appear before the Committee and offer my thoughts on the issues that need to be addressed in order to progress the Nation's high-level radioactive waste program. The opinions I will be presenting are based on my more than 40 years of experience working in the nuclear industry both domestically and internationally, and while serving as the Director of the Office of Civilian Radioactive Waste Management (OCRWM) in the Department of Energy from June 2006 to January 2009 in particular. My comments and opinions presented to the Committee are strictly my own and should in no way be construed as representing those of my current employer.

I appeared before this Committee in July 2006, almost eleven years ago, and committed to submit the License Application for the construction of the Yucca Mountain High-Level Radioactive Waste Repository by the end of June 2008. On June 3, 2008, I delivered the application to the offices of the Nuclear Regulatory Commission (NRC). That submittal started the formal licensing process for the repository as required by the Nuclear Waste Policy Act of 1982. In January 2015, the NRC staff issued the last of the five volumes of their Safety Analysis Reports (SERs) which marked the completion of their review of the application. They essentially found that the repository design meets all of the stringent design criteria, both during operation and after closure. Two issues were identified that would need to be addressed before the staff could recommend to the Commission that a construction license should be granted to the Department of Energy: 1) the Federal land on which the repository is to be sited has not been withdrawn permanently from public use by Congress and 2) the State of Nevada Water Engineer will not grant the Department the water withdrawal permits needed to build and operate the repository. Both of these issues must be addressed.

The next step in the licensing process that must proceed before the NRC can rule on the application is the adjudication of the numerous contentions that have been filed by several intervenors, the primary one of which is the State of Nevada. It is expected that the adversarial hearings which would take place before the Atomic Safety and Licensing Board (ASLB) would take two to three years to conclude before all of the contentions would be ruled upon. An appeals board would then hear any appeals by the parties. The final decision rests with the NRC commissioners. It will be at that point that we will know whether or not the Department will be allowed to build the repository at Yucca Mountain by the regulatory authority charged with making that decision.

So in terms of immediate next steps, the Department and the NRC would need to be funded to begin the hearing process to defend the License Application. The Department would need to re- assemble key members of its licensing team to write testimony and appear as experts witnesses. It would also need outside legal counsel who are experienced in nuclear regulatory law and litigation before the ASLB. A pre-requisite to all of this is that the Department must be an applicant that is willing and able to strongly defend its license application. That has not been the case over the past eight years.

Beyond the licensing process, there are issues which must be addressed by Congress in order to move forward with the Nation's high-level radioactive waste program, regardless of the outcome of the Yucca Mountain licensing process. These issues were identified and addressed in legislation that was proposed in 2006 and again in 2007 by the Bush Administration. That legislation was never acted upon and those issues are still relevant today and must be addressed by Congress. A brief summary of those issues follows.

Availability of the Nuclear Waste Fund

The ability of the Department to execute the long-term high-level radioactive waste program has been hampered, if not totally stopped, because it does not have access to funding from the Nuclear Waste Fund as was envisaged when the Fund was created by the Nuclear Waste Policy Act. In short, contributions to the Fund from the nuclear industry have been classified as Mandatory Receipts while the distributions from the Fund have been classified as Discretionary, subject to annual appropriations and budget scoring. This has had the effect of creating a varying and diminishing funding stream that makes it virtually impossible for the Department or any organization to execute a long-term capital program with any kind of schedule or cost certainty.

Permanent Land Withdrawal

As stated earlier in this testimony, the NRC staff found that without permanent withdrawal of the one hundred forty seven thousand (147, 000) acres of the Yucca Mountain site from future public use, the Department cannot demonstrate permanent control of the repository site. Congressional action is needed to withdraw the land.

Water Permits

As also stated earlier, the NRC staff found that the Department was unable to obtain, and the State of Nevada was unwilling to grant the necessary water withdrawal permits to the Department to allow construction and operation of the repository. Nevada has declared that the repository is not in the public interest and, therefore, will not grant the necessary permits.

Congress will need to declare that water use at and for the repository is in the public interest. It should also be noted that the Department has historically applied to the State of Nevada for Water Permits as a courtesy to show good faith and a willingness to engage with the State where Yucca Mountain is concerned. The Department does not apply for Water Permits at the Nevada National Security Site and could also by Administrative decision of the Secretary of Energy determine that they do not need to apply for Water Permits for Yucca Mountain. Either way, it is important to resolve this issue to the satisfaction of the Nuclear Regulatory Commission.

Transportation

Regardless of whether we will have a single central repository, a separate defense waste repository, or multiple interim storage sites, in all cases the spent nuclear fuel and high-level waste that currently reside at one hundred twenty one (121) sites in thirty nine (39) states need to be transported from those locations. There are hundreds of local jurisdictions that will be on those transportation routes and it reasonable to expect that some local and state authorities will attempt to block Department use of those transportation routes. Legislation is needed to clarify the Department's authority under the Atomic Energy Act to use federal Department of Transportation preemption if a local authority attempts to block a shipment.

Clarification of Federal Authority in Duplicative Regulatory Review Processes

There are several permitting actions that will be needed for the repository that are not under the jurisdiction of the NRC, but have been delegated to the states by various laws. For example, the Resource Conservation and Recovery Act (RCRA) provides the states regulatory oversight of waste types to be buried at disposal sites. Air emission permits are also administered at the state level. These permitting processes have the potential for additional local intervention and political influence that can stop repository construction and operation, even after the granting of the necessary licenses by the NRC. Legislation is needed to clarify federal authority over these permitting activities for the repository.

In conclusion, in order for the Country to move forward with the permanent disposal of its high- level radioactive waste and spent nuclear fuel, it needs three things: 1) a licensed place to put it, 2) the ability to move it from around the Country to that site, and 3) an organization that is adequately funded and has the requisite authorities so that it can be held accountable for the cost and schedule of executing the program in accordance with the law. Congress has the ability to address all three of these needs and it will need to do so in order for this national dilemma to be permanently solved. Technically, developing a repository is a fairly straightforward project. Politically, it is complex. If Congress can find a way to enable the project to move forward without political interference, the country will finally see success.

I look forward to answering any questions you might have regarding my testimony.

* * *

The Honorable Frank Pallone
Chairman, Committee on Energy and Commerce
2125 Rayburn House Office Building
United States House of Representatives
Washington, DC 20515

The Honorable Greg Walden
Ranking Member, Committee on Energy and Commerce
2322 Rayburn House Office Building
United States House of Representatives
Washington, DC 20515

Dear Chairman Pallone and Ranking Member Walden:

As your committee meets next week to discuss the future of high-level nuclear waste storage and disposal in the United States, I write to reaffirm the consistent position of the State of Nevada on the proposed Yucca Mountain Nuclear Waste Repository.

My position, and that of the State of Nevada, remains identical to the position of Nevada's past five governors: The State ofNevada opposes the project based on scientific, technical, and legal merits. I am totally opposed to any legislative effo1i to restmi the Yucca Mountain project. As you and your members know, under the Nuclear Waste Policy Act of1982, only the governor is empowered to consult with the federal government on matters related to the siting ofa nuclear waste repository.

My staff has thoroughly reviewed H.R. 2699, the proposed Nuclear Waste Policy Amendments Act of 2019. H.R. 2699 would do nothing to repair the central failing ofthe current federal law. In 1987, Congress substituted political science for emih science and selected Yucca Mountain in Nevada as the only site for repository development. H.R. 2699 would not only continue this failed policy; it would seriously weaken Nevada's cmTent due process rights to challenge documented safety concerns and adverse environmental impacts in the legally-mandated licensing proceeding.

This proposed legislation will waste billions of additional ratepayer and taxpayer dollars. Attempting to force an unsafe site on an unwilling state will fail. The proposed legislation only exacerbates the erosion of trust and confidence caused by the federal government's recent secret shipments of weapons-grade plutonium into our state.

I said in my State of the State address in Janum·y that not one ounce of nuclear waste will reach Yucca Mountain while I'm governor. I fully intend to keep my promise to the people of Nevada and fight against any attempts to restmi the failed Yucca Mountain program. If your committee is truly interested in fixing our nation's broken nuclear waste program, my staff and I, and Nevada's congressional delegation, would be happy to meet with you and explore constructive alternatives.

Respectfully,
Governor Steve Sisolak
State of Nevda
CC: Nevada Congressional Delegation
Members of the Committee on Energy and Commerce
State of New Mexico
Michelle Lujan Grisham
Governor

June 7, 2019
The Honorable Rick Perry
Secretary
U.S. Department of Energy
1000 Independence Avenue, SW
Washington, DC 20585

The Honorable Kristine Svinicki
Chairman
U.S. Nuclear Regulatory Commission
Mail Stop O-16B33
Washington, DC 20555-0001

Dear Secreta1y Perry and Chairn1an Svinicki:

I write to express my opposition to the proposed interim storage of high-level radioactive waste in the state of New Mexico. The interim storage of high-level radioactive waste poses significant and unacceptable risks to New Mexicans, our environment and our economy. Furthermore, the absence of a permanent high-level radioactive waste repository creates even higher levels of risk and uncertainty around any proposed interim storage site.

As you know, the Nuclear Regulatory Commission (NRC) is evaluating the issuance of a 40-year license to Holtec International for a consolidated interim storage facility in southeastern New Mexico. As proposed, this facility would store spent nuclear fuel (SNF) and reactor-related materials greater than low- level radioactive waste.

A facility of this nature poses an unacceptable risk to New Mexicans, who look to southeastern New Mexico as a driver of economic growth in our state. New Mexico's agricultural industry contributes approximately $3 billion per year to the state's economy, $300 million of which is generated in Lea and Eddy Counties, where the proposed facility is to be sited.

Further, the Pennian Basin, situated in west Texas and southeastern New Mexico, is the largest inland oil and gas reservoir and the most prolific oil and gas producing region in the world. New Mexico's oil and natural gas industiy contributed approximately $2 billion to the state last year. According to the U.S. Energy lnfonnation Administration (EIA), Lea County and Eddy County were ranked the second and sixth oil-producing counties in the country, respectively, earlier this year, with production continuing to mcrease.

Establishing an interim storage facility in this region would be economic malpractice. Any disruption of agricultural or oil and gas activities as a result of a perceived or actual incident would be catastrophic to New Mexico, and any steps toward siting such a project could cause a decrease in investment in two of our state's biggest industries. For those reasons, the New Mexico Cattle Growers' Association, the New Mexico Farm and Livestock Bureau and the Permian Basin Petroleum Association have all sent me letters opposing high-level waste storage in southeastern New Mexico. I have attached their letters for your review.

In addition to significant economic concerns about this project's potential impact on agriculture and the oil and gas industry, I am concerned about the financial burden it could place on the state and local conununities. Transporting material of this nature safely requires both well-maintained infrastructure and highly specialized emergency response equipment and personnel that can respond to an incident at the facility or on transit routes. The state of New Mexico cannot be expected to support these activities.

Finally, given that there is currently no permanent repository for high-level waste in the United States, any interim storage facility will be an indefinite storage facility. Over this time, it is likely that the casks storing SNF and high-level wastes will lose integrity and will require repackaging. Any repackaging of SNF and high-level wastes increases the risk of accidents and radiological health risks. Again, New Mexicans should not have to tolerate this risk.

Given the potential for adverse impacts to public health, the environment and our economy, I cannot support the interim storage of SNF.or high-level waste in New Mexico.

I thank you for your consideration of these concerns and look forward to your reply.

Michelle Lujan Grishnam
Governor
June 4, 2019

The Honorable Frank Pallone, Jr., Chairman
House Energy and Commerce Committee
2125 Rayburn House Office Building
Washington, DC 20515

The Honorable Greg Walden, Ranking Member
House Energy and Commerce Committee
2125 Rayburn House Office Building
Washington, DC 20515

RE: H.R.2699 - Nuclear Waste Policy Amendments Act of 2019

Dear Chainnan Pallone and Ranking Member Walden:

As the advocacy organization of Nevada's largest industry, representing 73 resorts statewide with a force of nearly 300,000 employees, the Nevada Resort Association stands vehemently opposed to the proposed nuclear waste depository at Yucca Mountain a position we have staunchly held since 1991.

Reviving this antiquated proposal has the potential to negatively impact tens of millions of citizens as transporting nuclear waste to Yucca Mountain for storage remains a dangerous and unstable proposal not only for Nevada but for every state these radioactive materials would travel through along the way. The potential for a massive disaster is simply too great to dismiss.

Yucca Mountain is located approximately 90 miles from Las Vegas, Nevada, one of the world's premier tourist, convention and entertainment destinations welcoming more than 42 million visitors annually and home to more than 2.2 million residents and still growing. In addition to being a leading global tourism destination, Nevada has returned to being one of the nation's fastest growing states given our strong economy, job growth and quality of life. Factors that would be irreparably damaged should this proposal move forward.

Nevadans recognize tourism as the lifeblood of our economy, supporting 450,100 jobs and generating nearly $68 billion in economic output annually. Tourism also provides nearly 40 percent of the revenue to our state's general fund thereby sustaining critical services and programs across the state. Beyond the obvious health and safety implications, Nevadans grasp the dire consequences of a nuclear waste depository at Yucca Mountain given the direct threat it poses to our tourism industry and, therefore, the economic stability of all Nevadans today and for years to come. Any decline in visitation due to visitors' perception of the safety of the region would have immediate and long-tenn repercussions on our state's economy and future growth. As the gateway to the West for our international visitors, our neighboring states would also feel the economic impact of such a slowdown.

As you consider H.R.2699 - Nuclear Waste Policy Amendments Act of2019, we add our industry's voice to the growing chorus of opposition against this ill-conceived legislation and stand with the countless concerned citizens here in Nevada and throughout the many states this life-threatening hazard would pass. We are united with our state's business community, environmentalists, Tribal leaders, Nevada's Governor and our entire federal delegation in our opposition of any efforts to restart a project based on outdated and uncertain science.

We urge you to rethink this nearly 40-year-old policy and instead pursue a consent-based solution for the storage of nuclear waste that meets the demands of today's challenges for the safety and security of our nation.

Sincerely, Virginia Valentine
President & CEO
Nevada Resort Association
CC: Members of House Energy and Conunerce Conunittee

County of San Diego

GEOFF PATNOE
DIRECTOR
(619) 531-5202

OFFICE OF STRATEGY AND INTERGOVERNMENTAL AFFAIRS
1600 PACIFIC HIGHWAY, SUITE 296, SAN DIEGO, CA 92101-2437

June 11, 2019

Chairman Frank Pallone, Jr.
House Energy and Commerce Committee
2125 Rayburn House Office Building
Washington, DC 20515

Ranking Member Greg Walden
House Energy and Commerce Committee
2322 Rayburn House Office Building
Washington, DC 20515

Dear Chairman Pallone and Ranking Member Walden:

On behalf of the County of San Diego (County), I am writing in support of H.R. 2995, the Spent Fuel Prioritization Act, and request your support of the legislation. This legislation, sponsored by Congressman Mike Levin, would prioritize the removal of spent nuclear fuel from a nuclear reactor based on the size of population near the site and seismic hazard associated with the area.

The San Onofre Nuclear Generating Station (SONGS) has been operating in the northwest corner of San Diego County for more than 45 years. In 2013, the station ceased all operations. Today, years after decommissioning began, approximately 1,700 tons of spent nuclear fuel remains stored onsite at SONGS. More than eight million people live within 50 miles of SONGS, which is 200 yards from the Pacific Ocean and 250 yards from both a major rail line and Interstate 5.

By prioritizing the removal of spent nuclear fuel from decommissioned nuclear sites in areas with large populations and high seismic hazards, H.R. 2995 is an important step toward a long-term solution for the removal of spent nuclear fuel around the country.

The County of San Diego supports H.R. 2995 and thanks you for your consideration of this legislation. Please do not hesitate to contact me if the County can provide any additional information on this issue.

Sincerely,

GEOFF PATNOE
Director
Office of Strategy and Intergovernmental Affairs
County of San Diego

Principal Man Ian Zabarte
Western Bands of the Shoshone Nation of Indians
P.O. Box 46301, Las Vegas, NV 89114

June 6, 2019

Chairman Frank Pallone
Committee on Energy and Commerce
2125 Rayburn House Office Building
Washington, DC 20515

For the Record: Nuclear Waste Policy Amendments Act of 2019—HR 2699.

Dear Chairman Pallone,

The Western Bands of the Shoshone Nation of Indians are an intervenor in licensing of the proposed Yucca Mountain high-level nuclear waste repository before the NRC Atomic Safety Licensing Board Panel Docket 63-001. Our primary contention is ownership. Site ownership is required by 10 CFR 63.121. Additional contentions cover ownership of water rights, and NEPA contention of disproportionate burden of risk to Native Americans and inappropriate radiation protection standards for Native Americans living a different lifestyle.

The DOE has failed to prove ownership that is vested in the Western Bands of the Shoshone Nation of Indians under the 1863 Treaty of Ruby Valley, Consolidated Treaty Series Volume 127 (1863) that is in *"full force and effect."* The Treaty of Ruby Valley is the *"supreme law of the land"* under the US Constitution and legislation should reflect the fact that Western Shoshone title remains unextinguished and the NRC does not adjudicate title.

We strongly oppose an amendment to the Nuclear Waste Policy Act in 2019 and look forward to making good on our ownership contention.

Sincerely,

Principal Man Ian Zabarte
Western Bands of the Shoshone Nation of Indians

Janet Tauro
Clean Water Action
198 Brighton Avenue
Long Branch, NJ

The Honorable Frank Pallone Jr.
United States House of Representatives
Washington D.C. 20510

June 11, 2019

Dear Congressman Pallone,

Thank you for receiving this letter for the record concerning the transportation of highly radioactive nuclear waste to a consolidated interim storage facility that is currently under review by the Environment and Climate Change Subcommittee, which will hear testimony tomorrow during its "Clean

Up Communities: Ensuring Safe Storage and Disposal of Spent Nuclear Fuel" hearing.

I have the honor to serve as the NJ Board Chair of Clean Water Action, but also live with my husband and have raised our family within the 50-mile ingestion zone of the Oyster Creek Generating Station in Lacey Township.

The facility is now closed. A sale and license transfer, which would include the $980 million decommissioning fund, is before for the federal Nuclear Regulation Commission (NRC).

Clean Water Action has been involved in identifying and reporting various safety concerns at the nuclear plant for almost two decades. Of particular concern has been the corrosion and deterioration of the plant from age and the accumulation of nuclear waste that will remain highly radioactive for thousands of years. There is about 1.2 million pounds of highly radioactive waste at Oyster Creek.

A permanent solution to nuclear waste storage has never been found and is unlikely to occur in the near future. We are left with choosing a least bad option. We agree with our colleagues at Beyond Nuclear that Hardened On-Site Storage (HOSS) to higher ground away from rising seas and worsening storm surges should be seriously considered for coastal areas until a permanent repository is established. Moving the waste thousands of miles out West to a temporary facility from which it would have to be moved again doubles the risk of a catastrophic accident.

How would the waste be moved out of NJ: by barge through Barnegat Bay and out to the Atlantic Ocean, up busy and congested Route 9 to the Garden State Parkway, or by rail through densely populated areas? The population here at the Jersey Shore has exploded; 650,000 permanent residents live in Ocean County and the number swells from tourism during the summer months to 2 million. About 3.5 million live within 50 miles of Oyster Creek.

The current plan at Oyster Creek calls for about 30 casks to be lined up like bowling pins on a pad area near Route 9. Requiring hardened storage, which creates a berm around the storage casks concealing them from terrorist attack, makes sense and increases public safety.

Maximizing safety must be the focus of the committee particularly since the NRC has allowed Exelon, or any future owner of Oyster Creek, to discontinue emergency planning around the plant once the fuel pool is emptied. In fact, Exelon tested its warning sirens for the last time two weeks ago. It would behoove this committee to also determine whether the company has been given the approval by NRC to disband its fire brigade, leaving a nuclear fire to local volunteer fire departments unequipped and untrained to handle such a disaster.

As I write this, my immediate and selfish inclination is to want that atomic waste packed up, shipped out, and away from my home. But, then, on reflection I ask if it is right to dump our poisonous garbage onto someone else.

Thank you for taking up this critical issue.

Sincerely,
Janet Tauro

The Honorable Paul Tonko
Chairman
Subcommittee on Environment and Climate Change
House Committee on Energy & Commerce
Washington, DC 20515

The Honorable John Shimkus
Ranking Member
Subcommittee on Environment and Climate Change
House Committee on Energy & Commerce
Washington, DC 20515

RE: "Cleaning Up Communities: Ensuring Safe Storage and Disposal of Spent Nuclear Fuel" Hearing

Dear Chairman Tonko and Ranking Member Shimkus:

On behalf of the National Association of Regulatory Utility Commissioners (NARUC), I want to thank the Committee for holding a hearing about "*Cleaning Up Communities: Ensuring Safe Storage and Disposal of Spent Nuclear Fuel.*" We respectfully request that this letter be included in the hearing record. As you consider the legislation before you today, we ask that you and your Subcommittee keep in mind the more than $40 billion in direct payments and interest the country's utility ratepayers have contributed to support the federal nuclear waste disposal program. NARUC would also like to applaud Representative McNerney and Ranking Member Shimkus for introducing the "Nuclear Waste Policy Amendments Act of 2019" (H.R. 2699). NARUC supports H.R. 2699, as introduced, just as we supported H.R. 3053 last Congress.

NARUC has previously expressed our views on the "Nuclear Waste Policy Amendments Act of 2019" discussion draft at a hearing before the Senate Committee on Environment and Public Works on May 1, 2019. H.R. 2699 is substantially similar to the Senate discussion draft and our views and positions on both pieces of legislation are consistent. Additionally, we have testified before this Subcommittee on nuclear waste disposal on numerous occasions in previous Congressional sessions, therefore we will not repeat those positions here.

Regarding H.R. 3136, NARUC believes the bill as it is currently drafted, while well intentioned, will do nothing to advance the nation's nuclear waste disposal program. Simply put, without conditioning interim storage on concluding the Nuclear Regulatory Commission's licensing process for a permanent repository, waste will not move. The fact is there will be difficulty with locating an interim storage site unless there is some permanent storage solution on the horizon. The Governor of New Mexico, whose recent pronouncement of opposition to an interim storage site in that State,[15] is a case in point.

Finally, NARUC has concerns with H.R. 2995. First, it appears that this legislation attempts to alleviate the waste issue for one State at the expense of the other 38 States that currently store waste on an "interim" basis. Second, with respect to decommissioned unit storage, we believe that the site which was decommissioned first, be the first to have waste removed followed by the next in the queue and so on. This is the only fair way to provide all the ratepayers with what they have paid for.

Thank you for your efforts to solve the nuclear waste disposal issue and NARUC looks forward to working with you and your staff going forward.

Sincerely,
Greg White

June 13, 2019

The Honorable Paul Tonko

---

[15] *See,* Lobsenz, George, *New Mexico, Texas governors move against nuclear waste facilities* (Energy Daily, June 11, 2019) (New Mexico Gov. . . . formally opposed the interim spent nuclear fuel storage facility proposed by Holtec . . . saying it poses "unacceptable risks" to the state because it threatens its agriculture and oil and gas industries and could become an "indefinite" storage site given the lack of federal action on a permanent high-level radioactive waste disposal repository."

Chairman
Subcommittee on Environment and Climate Change
U.S. House of Representatives
Washington, DC 20515

The Honorable John Shimkus
Ranking Member
Subcommittee on Environment and Climate Change
U.S. House of Representatives
Washington, DC 20515

Dear Chairman Tonko and Ranking Member Shimkus:

The American Gaming Association (AGA) appreciates you holding today's hearing on "Cleaning Up Communities: Ensuring Safe Storage And Disposal Of Spent Nuclear Fuel." As you explore potential solutions to address this important problem, AGA would like to express our vehement opposition to any effort to revive licensing activities for Yucca Mountain as a nuclear waste repository, including H.R.2699, the "Nuclear Waste Policy Amendments Act of 2019."

The AGA is the premier national trade group representing the $261 billion U.S. casino industry, which supports 1.8 million jobs nationwide. Our members include commercial and tribal casino operators, gaming suppliers and other entities affiliated with the thriving gaming industry in Nevada and across the country.

By reviving licensing activities for Yucca Mountain as a nuclear waste repository, this legislation has the potential to adversely impact citizens and businesses located in Nevada. That is why I and Chief Executives from 11 gaming companies and Nevada business organizations wrote to the full Congress in April opposing the Administration's budget request for Yucca funding.

Yucca Mountain is located just 90 miles from the world's premier tourist, convention and entertainment destination in Las Vegas, Nevada, which welcomed more than 42 million visitors last year. Las Vegas is once again on pace to meet or break that number with nearly 14 million visitors already accounted for in 2019. The Greater Las Vegas area is one of the fastest growing in the U.S. with a population that now exceeds 2.1 million people according to an estimate from the U.S. Census Bureau.

Safety and security remain a top priority for all Americans and any problems with the transport nuclear waste to the site throughout the country, or issues with its storage there, would bring potentially devastating consequences to the local, state and national communities. Moreover, with taxes on Nevada's tourism industry providing 42 percent of the state general fund, even a modest decline in visitors' perception about the region could have severe negative implications for the state's economy and future growth.

We strongly urge members to reject this flawed legislation and, instead, explore alternative solutions that respect state sovereignty and do not put Nevada's citizens and economy at risk.

We appreciate your attention to the concerns of citizens, the business community and Members of Congress from Nevada by ensuring radioactive waste is never stored anywhere near the world's leading tourist, convention and entertainment destination.

Sincerely,

William C. Miller, Jr.
President and Chief Executive Officer

Cc: The Honorable Frank Pallone, Chairman, Committee on Energy and Commerce

The Honorable Greg Walden, Ranking Member, Committee on Energy and Commerce

Members of the Committee on Energy and Commerce

Enclosure: Coalition Letter to Congress Regarding Yucca Funding from April 15, 2019

April 16, 2019

Dear Members of the U.S. House of Representatives and U.S. Senate:

As leaders in Nevada's entertainment, gaming, and resort industries, we write to express our vehement opposition to the inclusion of funding for the Yucca Mountain nuclear waste repository in the Administration's Fiscal Year 2020 budget request.

We urge you to continue working together to ensure that Yucca Mountain remains part of Nevada's past and that nuclear waste is never stored anywhere near the world's entertainment capital and Nevada's treasured public lands.

Yucca Mountain is located approximately 90 miles from the world's premier tourist, convention, and entertainment destination in Las Vegas, Nevada. In 2018, we welcomed more than 42 million visitors and we are on pace to meet or break that figure in 2019. The Las Vegas Valley is also home to 2.2 million residents, which includes hundreds of thousands of our employees, making it one of the largest and most economically dynamic cities in the country. The combination of these factors has a profound impact on the amount of revenue generated for Nevada's general fund. The impacts nuclear waste could have on our visitors and our employees would unquestionably have severe negative implications for Nevada's future and economic growth.

We are proud to stand together with Nevadans in every corner of the Silver State in opposition to this relic of the past and any attempt to resurrect a project that would so directly imperil the health, safety, and economic future of our great state. Thank you for your continued attention to this issue and please feel free to contact any of us should we be of any assistance to you on this matter.

Sincerely,

William C. Miller, Jr.
President and Chief Executive Officer
American Gaming Association

James Murren
Chairman and Chief Executive Officer
MGM Resorts International

Timothy J. Wilmott Chief
Executive Officer Penn National Gaming

Mark P. Frissora
President and Chief Executive Officer
Caesars Entertainment

Matt Maddox
Chief Executive Officer and President
Wynn Resorts, Limited

Virginia Valentine President
Nevada Resort Association

Sheldon Adelson
Chairman and Chief Executive Officer
Las Vegas Sands Corporation

Mary Beth Sewald
President and Chief Executive Officer
Las Vegas Metro Chamber of Commerce

Keith Smith
President and Chief Executive Officer
Boyd Gaming Corporation

Richard J. Haskins President
Red Rock Resorts

Joseph Asher
Chief Executive Officer
William Hill U.S.

Steven D. Hill
President and Chief Executive Officer
Las Vegas Convention & Visitors
Authority

* * *

June 12, 2019
The Honorable Frank Pallone
Chairman
House Committee on Energy & Commerce
U.S. House of Representatives
Washington, DC 20515

The Honorable Paul Tonko
Chairman
Subcommittee on Environment
and Climate Change
U.S. House of Representatives
Washington, DC 20515

The Honorable Greg Walden
Ranking Member
House Committee on Energy & Commerce

U.S. House of Representatives
Washington, DC 20515

The Honorable John Shimkus
Ranking Member
Subcommittee on Environment and Climate Change
U.S. House of Representatives
Washington, DC 20515

Dear Chairman Pallone, Ranking Member Greg Walden, Subcommittee Chairman Tonko and Subcommittee Ranking Member Shimkus:

Safe Energy Rights Group is writing today in regard to the June 13, 2019 Energy & Commerce Committee Sub-Committee on Environment and Climate Change (the "Committee") hearing "Cleaning Up Communities: Ensuring Safe Storage and Disposal of Spent Nuclear Fuel" (the "Hearing").

During their remarks at a Congressional Briefing on May 13, 2019 *(see* Environmental and Energy Study Institute website at: https://www.eesi.org/briefings/view/051319nuclear) former Nuclear Regulatory Commission ("NRC") Chairman Gregory Jaczko and retired Rear Admiral Len Hering both cautioned against the use of any interim storage facility, stating that such facility would likely become a *de facto* permanent site.

In addition, Admiral Hering stated that, "Of the 10 clear requirements established under [NRC Regulatory Regulation Title X Part 72], the thin wall [Holtec] container *onfy* provides a surety of 1. And the system used to transport and load those containers into their storage has extremely high likelihood of scratching, denting, or gouging the wall of that container, which, from a metallurgical perspective, provides for the opportunity for severe corrosion problems, which ultimately result in a potential breach of that container ... [and] a former engineer revealed that, had it been known this potential existed, they would have never approved it. What that tells me, is they have put into place a system for the movement of a 54-ton container that they had not tested, evaluated, or, in fact, seen.

These thin wall containers have no internal monitoring and no capability to be currently offloaded or transported - a specific requirement of Title X."

So, the first problem that must be addressed is the vulnerabilities in the design and regulation of dry storage systems currently being used. To be at all suitable for transportation, highly radioactive, irradiated nuclear fuel waste ("nuclear fuel waste") storage systems must be designed, fabricated and maintained to prevent both short-term and long-term radioactive leaks. The NRC does not require this.

There may not be any truly safe method for storing nuclear fuel waste for the millennia during which it will be dangerous. However, all approved canisters/casks must at the very least:

- be capable of being inspected and being repaired if necessary;
- be made using the best available materials;
- incorporate continuous radioactive monitoring;
- be designed so that the nuclear fuel waste can be retrieved and transferred to other containers if necessary;
- and be based on the best available current technologies and procedures with proven capacities to protect the American public.

Strict standards for the cooling of the nuclear fuel waste prior to loading must also be developed, taking into consideration the differences between conventional fuel and 'high-burnup' fuel and the effects of both on cannister/ cask integrity. Those standards should be based on current expert research - not on the most profitable timetable for the plant operator or successor licensees.

The Indian Point nuclear plant, located in my congressional district #17, already has nuclear fuel waste stored in 32 Holtec 'HI-STORM 218' and 32 Holtec 'HI-STORM IP1' containers. This nuclear fuel waste must be transferred to a more adequate system as an initial step. Transportation without dealing with the container and fuel loading problems is potentially catastrophic. Until these and other problems are resolved, there can be no safe process for relocating commercial nuclear waste from reactor sites to any other site - much less to a 'temporary' site and then to a permanent one.

The Committee is urged to issue guidance requiring the NRC to ensure that spent nuclear fuel containers and loading processes meet the above standards and resolve the above issues. In addition, no waivers of these essential safety standards should be allowed.

We hope that you will enter these comments into the Hearing record. Thank you in advance for your consideration to this request.

Regards,

Nancy S. Vann, President
Safe Energy Rights Group, Inc.

cc: Members of the House Energy & Commerce Committee, Subcommittee on Environment and Climate Change

Representative Paul Tonko, Chair
Representative John Shimkus, Ranking Member
Environment and Climate Change Subcommittee of Energy and Commerce Committee
US House of Representatives
2125 Rayburn House Office Building
Washington, DC 20515
June 13, 2019

Dear Representatives Tonko and Shimkus:

Nuclear Information and Resource Service (NIRS) is a national organization that serves as the information and networking hub for people and organizations concerned about nuclear power, radioactive waste, radiation, and sustainable energy issues since 1978. We submit these comments and attachments for the Hearing on "Cleaning Up Communities: Ensuring Safe Storage and Disposal of Spent Nuclear Fuel."

We oppose HR 2699, Nuclear Waste Policy Amendments Act of 2019, because it restarts licensing of the Yucca Mountain site which is fatally flawed technically, politically and morally. The site is seismically active, surrounded by volcanoes and unable to isolate the waste for the million years required by federal regulation. It is on sacred Western Shoshone land and violates principles of sovereignty, environmental justice and the Ruby Valley Treaty of 1863. The regulations had to be changed at least 4 times to prevent the site from being disqualified. Reopening Yucca siting is throwing good money after bad, wasting taxpayer money and further delaying responsible long term management of the nuclear power waste in the United States.

We oppose HR 2699 and HR 3136 because they legalize Consolidated "Interim" Storage (CIS, also termed MRS Monitored Retrievable Storage) of irradiated nuclear fuel. The reason Congress deemed CIS/MRS sites illegal is that they would likely become de-facto permanent without meeting conditions for permanent waste isolation. During the few years that Congress did consider such a site, it was clearly linked to the existence of an operating permanent repository, for which neither HR 2699 nor HR 3136 provide. These bills create *additional* radioactive waste sites and spread nuclear waste across the country.

Both provisions: resuming the failed Yucca Mountain licensing and legalizing CIS will lead to thousands of high level radioactive waste shipments on roads, rails and barges across the country, regularly for 40 to 50 years, and longer in the case of CIS. The routes go through 87%--386 of the 435 Congressional districts including 44 states and the District of Columbia, right through Capitol Hill, by Union Station alongside the Metro tracks.

Reactor communities deserve better, safer storage of nuclear waste until real permanent isolation can be found. Public interest groups have been calling for hardened onsite (or near site) storage with greater local control of decommissioning and waste management until there is permanent isolation. Further, the public deserves guaranteed inspections, maintenance, monitoring and repair all nuclear waste containers, the highest standards of independent nuclear quality assurance and continuous radiation monitoring systems, including on-line, real-time radiation monitoring.

If Congress changes the existing Nuclear Waste Policy Act, it should move forward, not backwards repeating past mistakes such as forced licensing of the cancelled Yucca site and reviving the defeated, unsuccessful consolidated "interim" storage program that expired in 1990.

Please see the two attached background documents on the problems with both Yucca Mountain in Brief and Consolidated "Interim" Storage Risks.

No false solutions to nuclear waste. Do not support HR 2699 or HR 3136.

Sincerely,

Tim Judson
Executive Director
Nuclear Information and Resource Service
timj@nirs.org

Attachment 1: Nuclear Information and Resource Service 6-13-19 Ltr to House Environment and Climate Change Subcommittee

Yucca Mountain in Brief

Nuclear weapons production and nuclear power reactor operation for electricity produce significant amounts of high level and other nuclear waste. To protect life on earth, as we know it, these materials must be isolated from the biosphere, essentially forever. In 2002, the plan to isolate these materials at Yucca Mountain for ten thousand years was judged to be inadequate – the D.C. District Court ruled that standards need to address a one million year time frame! In addition to being in a high risk earthquake area with volcanoes the Yucca site violates environmental justice, the Ruby Valley Treaty of 1863 and Native American sovereignty.

Nuclear waste is an issue that needs urgent attention. However, flawed, temporary and ad hoc plans endanger current and future generations. The highest scientific and technical specifications must be rigorously followed to ensure long-term protection of public health and wellbeing.

With the opening of a new session of Congress, there is yet another ill-conceived effort to revive the proposed and administratively cancelled Yucca Mountain high-level nuclear waste repository program, which has been dormant since 2010 – the site is undeveloped except for an entrance tunnel and is not appropriate for receiving nuclear waste. It is useful to review elements of the long history of the Yucca project that have led to its current state of near demise and assess whether the national interest in the safe, long term isolation of commercial and weapons related high-level nuclear waste would best be served by officially abandoning the project as irrevocably flawed and seek a fresh, consent-based and technically rigorous start at meeting the nation's need for safe management and disposal of these long-lasting, highly dangerous wastes.

Throughout this process, there have been repeated efforts to consolidate high level waste prior to having an operating permanent place to isolate it. These proposals consistently and rightly have been rejected.

It is now just over 31 years since Congress passed the 1987 Nuclear Waste Policy Amendments Act that terminated the deliberative national nuclear waste repository site screening process, and instead, made decisions based on political power rather than sound science. Thus, the Yucca Mountain, Nevada candidate site became the only site for potential repository development. Five years later, in 1992, Congress instructed the Environmental Protection Agency to craft a radiation protection standard specific to the Yucca Mountain site *because it was generally agreed by an EPA Science Advisory Board panel that the site could not meet the "generally applicable" safety standard required by the 1982 Nuclear Waste Policy Act for judging the adequacy of any potential repository site*. The Nuclear Regulatory Commission was instructed to conform its repository licensing rules to the new EPA Yucca Mountain specific standard.

The 1987 Amendments Act retained the original Act's requirement that the Department of Energy establish Site Recommendation Guidelines that among other things define factors to qualify or disqualify potential repository sites' consideration as a candidate site. Just prior to the Secretary of Energy's

2002 Site Recommendation of Yucca Mountain to the President, the Department of Energy amended the Site Recommendation Guidelines to eliminate those factors that would qualify or disqualify a site and thus preserved Yucca Mountain as the only site to be considered for repository development. Nevada governors twice, first in 1989 and again in 1999, had informed the Secretary of Energy that the site should be disqualified because of substantial evidence that the site did not meet the required minimum groundwater travel time from the repository waste-emplacement-zone to the accessible environment, a factor critical to the site's ability to isolate radioactive waste. The response to both governors' letters was that DOE was still studying the site. DOE's ultimate response was to repeal the guidelines that Yucca Mountain could not meet and would make the site unsuitable for further consideration.

Nevada's statutorily permitted Notice of Disapproval of the President's recommendation of the Yucca Mountain site to Congress was overridden by the House and Senate in 2002. According to the Act, DOE then had up to 90 days to submit a Yucca Mountain repository site license application to the Nuclear Regulatory Commission. DOE intentionally ignored this requirement in its rush to a Site Recommendation, and it was not until June 2008, six years later, that DOE submitted its Yucca Mountain repository license application to NRC, months prior to promulgation of a final radiation protection standard, which prompted a 2009 revision of the application.

Nevada timely filed a petition to intervene in the NRC licensing proceeding and had 218 contentions admitted for adjudication. Other intervening parties brought the total contentions to about 300 – by far a record for any NRC adjudicatory proceeding. In its early substantive review of the license application, NRC Staff issued 642 Requests for Additional Information, 50 of which resulted in DOE's commitment to update its license application, which has not taken place since the licensing proceeding was suspended in 2010, and remains suspended today, for lack of appropriations from Congress. Prior to the 2010 suspension, DOE had unsuccessfully moved to withdraw its license application as being "unworkable" for reasons later specified to include the unrelenting resistance to the program by Nevada. Nevada leadership has communicated numerous times that it does not consent to a Yucca Mountain nuclear waste repository, and vows that it will not consent to it.

Following the suspension of the proceeding, the States of Washington and South Carolina, and Aiken, South Carolina brought NRC into the Circuit Court of Appeals for the District of Columbia by claiming the suspension was unlawful. The court ordered NRC to continue the licensing process as long as it had unexpended appropriated funds from prior years, but it could not order NRC to proceed if no further funds were appropriated. To date NRC has spent nearly all of its carry-over funds writing a needed Supplemental Environmental Impact Statement on groundwater impacts and a Safety Evaluation Report that presents the NRC Staff position on the adequacy of the license application to provide "reasonable expectation" that the repository will meet the specifically tailored radiation protection standard for Yucca Mountain. The Staff concluded that its position is, should there be an adjudication of the application, that it meets the required safety standard, but it lacks the required Congressional land withdrawal for the site, and an appropriation of water rights, which Nevada had earlier denied.

Aside from the 50 commitments to update the license application noted above, a massive revision of the license application is necessary because the repository design and safety analysis is based on a disposal waste package concept that DOE has abandoned due to the evolution of waste storage technology at nuclear reactor sites during the period of dormancy of the project. The plan in the application was to have the waste removed from reactor cooling pools and placed directly into an industry-wide uniform canister whose specifications are the basis of the design and safety analysis of the Yucca Mountain repository. The canisters would then be part of an integrated system of storage, transportation, and disposal. The design and regulatory certification work for the canister was abandoned by DOE shortly after the licensing process was suspended, and in the meantime reactor owners have been purchasing, and will continue to purchase, a wide variety of containers for at-reactor storage of their irradiated ("spent" or "used") nuclear fuel, none of which even approach the specifications of the planned canister consistent with the repository design. Repackaging the irradiated fuel into canisters meeting the original specifications has significant time, expense, and worker exposure issues that will continue to grow as irradiated fuel is discharged from operating reactors. It is unknown what path DOE would take to proposing a waste package for disposal at Yucca Mountain.

Because the fractured characteristic of the rock at Yucca Mountain provides pathways for infiltration of precipitation water through the underground location of the waste, DOE has designed a relatively corrosion resistant disposal container for the planned canister. But it is known that the container will eventually fail and infiltrating water will carry the waste radionuclides to the water table and the accessible environment. In an attempt to further delay release of the waste to the environment, DOE has planned installation of 11,500 titanium drip shields over the containers to deflect water that would drip onto the containers. The ability to perfectly install the 5 ton drip shields with complex waterproof interlocking joints using robots in a high heat and radiation field has not been demonstrated and may not be possible if there is rock fall in the unmaintainable tunnel. Installation of the drip shields is planned for the final ten years of the 100 year operating lifetime of the repository at a cost estimated at $5 to $9 billion in 2018 dollars. Of course, if the statutory capacity of the repository is expanded, as proposed in H.R. 2699, or eliminated entirely, the drip shield cost increase could more than double. *The question for Congress today is, "Are you willing to commit the country to an expenditure of multi billions of dollars at least 100 years from now on an unproven technology after all the waste has been emplaced underground?" Such a commitment is necessary because without the drip shield, DOE's own analysis formula shows that the radiation protection standard will be violated, and the site would be unsuitable for disposal.*

If there is intent to revive the Yucca Mountain repository project some near-term commitments are required. At least $2 billion will be required to complete the licensing process over the next few years, with no assurance of how long it will take. If a license is approved, it assuredly will be followed by extensive litigation. A new 300 plus mile rail line to Yucca Mountain must be approved and constructed, and a portion realigned from the original plan because of a recent National Monument designation. Initial repository (without rail line) construction costs would require appropriations of over $1 billion (2008 dollars) per year. Complete construction of the statutory 70,000 metric ton repository would require boring at least 40 miles of new tunnel at Yucca Mountain.

The Yucca Mountain project has a unique and troubling history; the geology of the site is known to not be capable of isolating nuclear waste; the 2008 license application repository design and safety analysis requires significant revision; and the safety of the site remains to be heavily contested in an unprecedented licensing hearing of uncertain outcome. It is time to abandon this failed Yucca Mountain nuclear waste repository project and open the way for a new nuclear waste policy.

Attachment 2 NIRS 6-13-19 Letter to House Environment and Climate Change Subcommittee

Consolidated "Interim" Storage of High-Level Radioactive Waste-Prevent a Disaster by Rejecting This Risky Plan

We oppose any bills that authorize, fund, facilitate or enable consolidated 'interim' storage of nuclear reactor waste, mainly irradiated ("spent") nuclear fuel.

Consolidated "Interim" Storage entails moving and storing high-level radioactive waste, one of the most deadly substances on Earth, from dozens of nuclear sites to one or more centralized locations, presumably until it can be sent to a permanent repository. (The cancelled Yucca Mountain site will not serve this purpose.) A dangerous unprecedented program to haul over a hundred thousand tons of irradiated nuclear fuel thousands of miles across the country, only to move it again, is counterproductive and senseless. It means creating one or more new radioactive sites, in addition to the ones already contaminated.

Thousands of radioactive shipments would travel routinely for 20 to 40 years through 87% of Congressional districts, emitting radiation and creating transportation accident and security risks all the way, putting nearly the whole country at risk. Meanwhile, more waste is still being generated and stored at operating nuclear reactors.

The supposedly "interim" site(s) could become *de facto* permanent even though it would not be designed or characterized for it. Consolidation is the first step to dangerous reprocessing of nuclear waste and plutonium proliferation.

Consolidated "Interim" Storage would not facilitate or accelerate the transfer of waste to a permanent repository; in fact it would delay and take resources away from efforts to do so. It would cost billions of dollars, money better spent on a realistic scientific effort to develop viable permanent isolation and improved storage systems.

Two companies have applied for licenses from the Nuclear Regulatory Commission (NRC) for consolidated (centralized) "interim" storage of high-level radioactive waste. Both applications are being challenged on legal and technical bases.

- Waste Control Specialists (WCS) seeks to bring 40,000 tons of high-level radioactive waste to Andrews County, Texas, from nuclear reactors around the country. WCS and ORANO USA have formed Interim Storage Partners. ORANO is the French government-owned corporation formerly known as Areva, which has been cited for failure to perform on a number of Department of Energy contracts. Their application seeks storage for 40 years but anticipates extensions up to 100 years.
  The above–ground dry storage casks would be exposed to extreme desert temperatures, storms, lightning, flooding and seismic events. The site is close to the nation's largest aquifer, the Ogallala, which provides water for eight states including the nation's bread basket.
- Holtec International seeks to store up to 173,000 tons of deadly high-level radioactive waste for 40 years, with possible extensions to 120 years, at a site between Hobbs and Carlsbad, New Mexico. The tops of storage units would be slightly above the ground surface, with waste canisters below ground, at a site with groundwater present at depths of 35' to 50' below the surface. This site would have similar risks from extreme temperatures, intense storms and earthquakes.

Significant Risks

You can't see, taste, smell or feel it, but ionizing radiation can lead to birth defects, cancers, reduced immunity and death. Exposure to unshielded high-level radioactive waste is lethal. Accidents involving radiation releases can lead to water contamination and cost taxpayers billions of dollars for cleanup. One radioactive transport accident could destroy lives and livelihoods and impact water supplies, businesses, homes, ranches, agriculture, thriving local industries including the oil industry, and tourism. Rails, trucks, and barges could all be used to transport this deadly waste.

Over 10,000 shipments would take place, in a process lasting over 20-40 years, risking lives and financial disaster. Rail shipments would involve very heavy loads, weighing as much as 38% more than train tracks are rated to handle. Real-world train accidents have already exceeded the supposed worst-case scenarios used for analyzing risks, including a head-on collision of two trains in West Texas, each going 65 miles per hour. However, the NRC is not requiring updated standards that would meet or exceed the severity of accidents that have already happened, nor is the potential use of drones and armor-piercing weaponry in sabotage events being considered.

Consolidated "interim" storage plans pose risks for the entire nation, since transportation routes would go through many major U.S. cities, as well as rural areas across the country. Any major commercial rail line (including through the heart of Washington, D.C.) could be used for transport of high-level radioactive waste, and 218 million people live within a half-mile of likely rail lines, putting them at increased risk for exposure, even from routine emissions. Risks could escalate when a train is stopped at a siding or is in a switchyard.

A study by Radioactive Waste Management Associates in 2003 found that 1,370 latent cancer fatalities could result from a rail accident with irradiated fuel. They estimated costs of $145 - $270 billion[16] for a severe rail accident.

A single rail car could carry as much plutonium as was in the bomb dropped on Nagasaki. Transportation accidents and potential acts of terrorism could lead to radioactive contamination of land, water and air, and radioactive emissions from routine transport shipments could impact health and safety. These unnecessary risks should be prevented.

The communities and regions targeted with Consolidated "Interim" Storage don't want it, despite claims to the contrary. There is strong bipartisan opposition.

- Resolutions opposing consolidated interim storage of this waste and its transport through local communities have been passed by Bexar, Dallas, Nueces, El Paso and Midland Counties in Texas, the cities of San Antonio, Midland and Denton, and the Midland Chamber of Commerce. In New Mexico, resolutions have been passed in Bernalillo McKinley and Santa Fe Counties, the cities of Albuquerque, Las Cruces, Lake Arthur, Jal, Gallup, the Church Rock Chapter of the Navajo Nation and the New Mexico Cattle Growers Association.
- A coalition of oil and ranching companies and royalty owners are legally opposing the high-level radioactive plans for Texas and New Mexico, due to risks to the oil industry.
- Nine New Mexico Senators and twenty-one House Members asked the NRC to allow the Legislature time to examine crucial health, safety and economic concerns to the state.
- Together, over 70,000 public comments were submitted to the NRC opposing the two high-level radioactive waste projects. There is no consent to these plans. In fact, there is very strong opposition.

---

[16] In today's dollars, then it would be nearly 40% greater, with inflation.

- The New Mexico Governor has expressed opposition.

Please act in the interest of people across the nation in preventing the dangerous plans to transport massive amounts of high-level radioactive waste to consolidated "interim" storage sites.

* * *

June 13, 2019

Chairman Paul Tonko Subcommittee on Environment
and Climate Change
House Committee on Energy & Commerce
2125 Rayburn House Office Building
Washington, DC 20515

Ranking Member John Shimkus
Subcommittee on Environment and Climate Change
House Committee on Energy & Commerce
2322 Rayburn House Office Building
Washington, DC 20515

Dear Chairman Tonko and Ranking Member Shimkus:

Thank you for holding a hearing today on spent nuclear fuel. I am Arlen Orchard, CEO of the Sacramento Municipal Utility District (SMUD), which is a member of the Decommissioning Plant Coalition (DPC).[17]

By way of background, SMUD's Rancho Seco Nuclear Generating Station ceased commercial operation in 1989, decommissioning planning began in 1991, and commodity removal began in 1997. In October 2009 the Nuclear Regulatory Commission (NRC) released the majority of the site for unrestricted public use, and on August 31, 2018 the NRC terminated the Part 50 license. This leaves the approximately 14 acres of land licensed under 10

[17] The DPC was established in 2001 out of the recognition that the overwhelming attention of the regulator, the industry and policy makers would be focused on the operating fleet and provides a forum for the identification of federal policy and regulatory issues of unique or special concern to decommissioning civilian nuclear facilities. Since its inception, plants that have been represented in the work of the DPC include: Big Rock, Connecticut Yankee, Crystal River, Duane Arnold, Humboldt Bay, Kewaunee, LaCrosse, Maine Yankee, Pilgrim, Rancho Seco, San Onofre, Vermont Yankee, Yankee Rowe and Zion.

CFR Part 72 as SMUD's Rancho Seco Independent Spent Fuel Storage Installation (ISFSI), excluding approximately 11 acres of land that holds an Independent Spent Fuel Storage Installation (ISFSI) that contains 22 dual-purpose systems licensed for the dry storage and transportation of used nuclear fuel and Greater-Than-Class-C (GTCC) waste ultimately destined for disposal by the Department of Energy (DOE).

As is the case with other contract holders[18] SMUD has litigated a partial breach of contract claim against DOE, seeking to recover the costs incurred in our management of this material, which the Department was required to begin accepting in 1998. To date, SMUD has won judgments in the U.S. Court of Claims of more than $100 million, covering costs through June 2015. These monies have been paid out of a permanent appropriations account in the Treasury called the Judgment Fund.

From the outset, one of the chief goals of the DPC has been to hasten the day when the federal government will meet its contractual obligations to remove the used fuel and GTCC material stranded on our various sites. As the Nuclear Waste Policy Act (NWPA), as amended in 1987, was already 14 years old when the DPC formed, we supported the Yucca Mountain project and worked with Congress in urging DOE to prepare a sound license application, address the transportation infrastructure requirements necessary to support a phased-in shipping campaign, and otherwise take steps necessary to prepare for the movement of this material from our sites on a priority basis.

As I suspect is the case with other contract holders, we watched with concern the development of political opposition to the Yucca Mountain project in the State of Nevada and our concerns were realized when the Obama Administration determined that the project was no longer viable and began to close down the licensing effort beginning in 2009. Ten years later, we regret the political impasse over Yucca has kept spent nuclear fuel at Rancho Seco and other similar sites across the country.

[18] In the Nuclear Waste Policy Act of 1982 (NWPA), the owners of civilian nuclear power reactors were required to enter into contracts with the DOE and pay a fee, based on the amount of electricity generated at those reactors. Those fees have been deposited into the Nuclear Waste Fund, invested in interest bearing accounts and are to be made available for the siting, construction and operation of facilities described in the Act. In return, the DOE was obligated to begin accepting used fuel at each reactor, based generally on the concept of the oldest fuel first, in 1998. The NWPA and the Standard Contract developed pursuant to the Act (found at 10 CFR 961) allows the DOE to accord priority to any used fuel or GTCC waste "removed from a civilian nuclear power reactor that has reached the end of its useful life or has been shut down permanently for whatever reason."

SMUD and the DPC have supported the establishment of a voluntary, incentive-based siting program that would lead to the licensing of a consolidated interim storage (CIS) facility and a pilot program to remove the material from our sites on a priority basis. This pilot would demonstrate the ability of the federal government to plan and execute their responsibilities for used fuel and GTCC waste acceptance and transportation under the Standard Contract, relieve the taxpayer of the obligation to continue paying Judgment Fund damages and allow these sites to be freed for other useful purposes. We believe this intermediary path is a viable alternative to the status quo.

SMUD appreciated the efforts the Chairman and Ranking Member of this Subcommittee have made to address the siting of a permanent repository for spent nuclear fuel at Yucca or elsewhere. While we hope that the impasse can ultimately be resolved, we believe this committee should act to advance an interim storage option. We are particularly grateful for the efforts of Rep. Doris Matsui. Her bill – the STORE Nuclear Fuel Act – would make progress by:

- Allowing the Department of Energy (DOE) or a private entity to develop consolidated interim storage (CIS) locations through a consent-based process; and
- Permitting DOE to take title of SNF to transfer the material from their present locations to the CIS locations.

Mr. Chairman and Members of the Subcommittee, as you examine possible legislative options to address our current policy failure, and there really is no other word to describe the current situation, we urge you to include provisions that would support the establishment of a CIS program, one that can be accomplished with reasonable support from the Nuclear Waste Fund without any impact upon the repository program.

We believe that the inclusion of Consolidated Interim Storage will restore the confidence of our local communities in the federal government's will to meet its obligations and promises.

We believe that establishing a Consolidated Interim Storage program will address increasing regulatory costs at our sites as the material remains on site for longer periods than anyone ever imagined.

We believe that a successful Consolidated Interim Storage program will enable our communities to repurpose the multiple sites that are currently restricted by safety and security requirements.

We applaud your steadfast interest in a vibrant repository program, and we urge you to strongly consider Rep. Matsui's STORE Nuclear Fuel Act.

Thank you for the opportunity to provide this written testimony. SMUD and the DPC are always prepared to assist the committee as it considers this critical issue.

Sincerely,

Arlen Orchard
CEO
Sacramento Municipal Utility District

Beyond Nuclear
7304 Carroll Avenue, #182
Takoma Park, MD 20912
Web: www.beyondnuclear.org

June 13, 2019
Letter for the Record, by Kevin Kamps, Beyond Nuclear

Re: Energy & Commerce Committee/Environment & Climate Change Subcommittee, "Hearing on Cleaning Up Communities: Ensuring Safe Storage and Disposal of Spent Nuclear Fuel"

Dear Chairman Pallone, Ranking Member Walden, Chairman Tonko, and Ranking Member Shimkus,

Please accept this letter for the record. My name is Kevin Kamps. I serve as Radioactive Waste Specialist at Beyond Nuclear based in Takoma Park, MD. We object to H.R. 2699, the "Nuclear Waste Policy Amendments Act of 2019," just as we objected to last session's similar to identical H.R. 3053.

These dangerously bad bills advocate for the scientifically unsuitable Yucca Mountain dump, which is neither consent-based, nor environmentally just – nor even legal, given Western Shoshone Indian title to land and water, per the "peace and friendship" 1863 Treaty of Ruby Valley, signed by the U.S. government, the highest law of the land, equal in stature to the U.S. Constitution itself. In opposing the Yucca dump, we join with well over a thousand environmental and environmental justice organizations which have expressed similar opposition to the Yucca dump over the past 32 years, ever since the "Screw Nevada" bill of 1987.

We also oppose the "Spent Fuel Prioritization Act of 2019," H.R. 2995. Why would a relatively recently shut down nuclear power plant, like San Onofre in California, and even still operating atomic reactors, as at Diablo Canyon nuclear power plant in California, get to cut in line, in regards to high-level radioactive waste export shipments? Big Rock Point in Michigan, for example, has been shut down since 1997. Yankee Rowe in Massachusetts shut down in 1992. There are a dozen or two reactors in this country that shut down prior to San Onofre Units 2 and 3 in 2013. What ever happened to oldest fuel first?! California reactors should not be allowed to cut in line. They are not the only ones in the country at risk of earthquakes, tsunamis, rising seas, and other disaster risks.

As the environmental movement has called for since 2002, Hardened On-Site Storage (HOSS) of high-level radioactive waste should be implemented. If not on-site, then as near to the site of generation as is safely possible. For example, San Onofre's high-level radioactive wastes could be moved a few miles east, deeper into the heart of Camp Pendleton Marine Corps Base. This would remove the wastes from the earthquake fault line area, from the tsunami zone, away from rising seas – with the added bonus of thousands of U.S. Marines to help guard them. This, rather than ship San Onofre's waste a thousand miles to the east, to New Mexico and/or Texas, for "interim storage," from where they would someday (or some decade, or some century) have to be moved yet again – to where we know not – instantly doubling transport risks, for no good reason.

We also oppose the "STORE Nuclear Fuel Act of 2019," H.R. 3136. This bill would make legal the U.S. Department of Energy (DOE) taking title (ownership) of commercial irradiated nuclear fuel at a privately owned consolidated interim storage facility (CISF), such as those currently targeted at New Mexico by Holtec International/Eddy-Lea Energy Alliance, and at Texas by Waste Control Specialists/Interim Storage Partners. Besides the environmental injustice, or radioactive racism, inherent in targeting a majority

Hispanic region with the deadliest radioactive wastes in the U.S., there is also the pollution burden already borne by these communities, due to already present, intensive nuclear and fossil fuel industrial activities.

But in addition, as New Mexico Governor Michele Lujan Grisham wrote in a letter to Energy Secretary Perry, and Nuclear Regulatory Commission Chairman Sviniki, just last week, the lack of a permanent repository means that "consolidated interim" risks becoming *de facto* permanent, surface storage. The U.S. Department of Energy itself warned in its Feb. 2002 Final Environmental Impact Statement for the Yucca Mountain dump, that high-level radioactive waste abandoned at the surface of the land, given loss of institutional control, and inevitable degradation and failure of containers over long enough time periods, would result in catastrophic releases of hazardous radioactivity to the environment. This could happen on-site at reactors, but it could also happen away-from-reactor, as at CISFs in New Mexico and/or Texas. Due to such risks, the governor of New Mexico has communicated clearly to decision makers that her state does not consent to taking on such risks. Secretary Perry's statement in response to a question from U.S. Rep. Mike Simpson, at a House Budget Committee hearing in late March, that "interim" becoming permanent in west Texas would be fine by him, and by west Texans, shows that Secretary Perry is not even familiar with his own department's warnings about the high risks of containers failing in the future, causing catastrophic radioactivity releases downwind, downstream, up the food chain, and down the generations.

The common theme of all these bills, that would rush open a permanent repository at Yucca Mountain, and/or CISFs in New Mexico and/or Texas, is the transport risks.

Our country needs to avoid radioactive waste wrecks, both figurative – of policy – as well as literal, on our roads, rails, and waterways. We need to just say no to unwise irradiated nuclear fuel transport, storage, and disposal schemes, that have more to do with offloading nuclear utilities' liabilities onto the public, than on protecting health, safety, and the environment. Transporting high-level radioactive waste by truck, train, and barge, through 44 states and the District of Columbia, to the unsuitable Yucca Mountain, Nevada site would take unnecessary risks, and violate consent-based and environmental justice principles.

Yucca is the worst site ever studied for high-level radioactive waste disposal. It has been kept alive by "double standard standards": when Yucca can't meet the standards, they are either weakened or gotten rid of. Yucca is an earthquake and volcanic zone. If radioactive waste is ever buried there, it will leak massively into the groundwater, creating a large nuclear sacrifice zone downstream. Nevada has not consented to being railroaded into becoming this country's radioactive waste dump. The Western Shoshone Indian Nation, who live downstream, have accused federal officials of environmental racism, and even genocide, from both nuclear weapons testing on their land, as well as proposals for high-level radioactive waste dumping.

Consolidated interim storage also makes no sense. Take Private Fuel Storage, targeted at the Skull Valley Goshutes Indian Reservation in Utah. If that *de facto* permanent surface "parking lot dump" had opened, and imported 4,000 Holtec casks of high-level radioactive waste, they would have been "returned to sender" when Yucca was wisely cancelled by the Obama administration in 2010. More than 50 casks from Maine Yankee would have travelled 5,000 miles round-trip, through a dozen states, for nothing. If private, consolidated interim storage is such a good idea, why then was the PFS license – rubber-stamped by NRC – never utilized? This clearly shows that Holtec, as well as WCS, are depending on DOE to pay all the bills for interim storage in NM and/or TX – which happens to be illegal, under the Nuclear Waste Policy Act of 1982, as Amended.

High-level radioactive waste shipments are potential Mobile Chernobyls. Risks include long-lasting, high-temperature fires, as the National Academies of Science acknowledged in 2006, which could breach shipping containers and release disastrous amounts of hazardous radioactivity in heavily populated areas. Barge shipments – on the Great Lakes, CA's Pacific coast, the waters of NJ, NY, and CT, and numerous other rivers, and sea coasts (including, shockingly, from Indian Point, down the Hudson, past Manhattan!) – are potential Floating Fukushimas, risking radioactive contamination of vital drinking water supplies and the food chain, and even deadly nuclear criticality accidents if submerged.

A quality assurance meltdown in industry and at NRC, revealed by whistleblowers and accidents, adds to the risks of shipments, calling into question – as but one example – Holtec casks' structural integrity sitting still, let alone travelling 60mph or faster on the rails. NAS also emphasized that risks of terrorist attack on shipments need to be addressed. A 1998 test of a TOW anti-tank missile on a high-level radioactive waste shipping container conducted at the U.S. Army's Aberdeen Proving Ground showed that casks are potential dirty bombs on wheels. Combined with an incendiary, such breaches could cause a large-scale radioactivity release.

Incredibly, DOE has thrown caution to the wind, conducting unprecedented LIQUID high-level radioactive waste truck shipments from

Chalk River, Ontario, to Savannah River Site, South Carolina, beginning in spring 2017, with little to no Environmental Assessment. DOE has also, more recently, secretively shipped ultra-hazardous weapons-grade plutonium from South Carolina, to Nevada, while in bad faith pretending to be litigating the matter in federal court, in the face of Nevada's non-consent. DOE is out of control, and cannot be trusted to undertake radioactive waste burial at Yucca Mountain, nor even serve as the primary or sole customer for high-level radioactive waste storage at CISFs in New Mexico and/or Texas (an activity barred by the Nuclear Waste Policy Act of 1982, as Amended).

Holtec is also not to be trusted. It was recently revealed by WNYC and ProPublica that Holtec lied on a tax break application form in New Jersey, regarding a ban on its doing business with the Tennessee Valley Authority in the aftermath of a bribery scandal. This has resulted in a grand jury investigation in New Jersey, as well as a hold put on Holtec's $260 million tax break.

Waste Control Specialists also cannot be trusted. Just last week, WCS lobbyists, and its allies in the Texas State Legislature, attached a rider to a domestic violence bill, in an effort to stop paying a 5% commission from revenues on low-level radioactive waste disposal in Texas, to the state government. Texas Governor Abbott vetoed the entire bill over the matter, a blow to domestic violence prevention and recovery services, due to WCS's reckless greed.

Thus, we urge opposition to H R. 2699, the "Nuclear Waste Policy Amendments Act of 2019," H.R. 2995, the "Spent Prioritization Act of 2019," and H.R. 3136, the "STORE Nuclear Fuel Act of 2019." Thank you for considering our views.

Sincerely,
Kevin Kamps
Radioactive Waste Specialist
Beyond Nuclear
(240) 462-3216
kevin@beyondnuclear.org

The mission of Beyond Nuclear is to educate and activate the public about the connections between nuclear power and nuclear weapons and the need to abolish both to safeguard our future. Beyond Nuclear advocates for an energy future that is sustainable, benign and democratic. The Beyond Nuclear team works with diverse partners and allies to provide the public, government officials, and the media with the critical information necessary to move humanity toward a world beyond nuclear. Beyond Nuclear is a 501(c)(3) nonprofit organization.

* * *

June 12, 2019

The Honorable Paul Tonko
Chairman
Environment & Climate Change Subcommittee
House Committee on Energy & Commerce
2125 Rayburn House Office Building
U.S. House of Representatives
Washington, D.C. 20515

The Honorable John Shimkus
Ranking Member
Environmental & Climate Change Subcommittee
House Committee on Energy & Commerce
2322 Rayburn House Office Building
U.S. House of Representatives
Washington, D.C. 20515

Dear Chairman Tonko and Ranking Member Shimkus:

Thank you for the opportunity to write you regarding legislation your subcommittee is considering to amend the Nuclear Waste Policy Act (NWPA). I am Leo Blundo, a member of the Nye County Commission and the designated liaison commissioner for nuclear waste issues. I was elected last year to my commission seat. I raise that issue to point out that the Nye County Commission remains in support of the proposition that the Yucca Mountain licensing proceeding should be completed so that the science behind the proposed repository can be fully explored and evaluated by qualified scientists and technical experts.

There has been lots of rhetoric about Yucca Mountain and many misleading statements put out about it. I, and my predecessors, have testified and written to this committee, and others, always in an attempt to keep the discussion on the facts and away from political and emotional side issues. We appreciate the opportunity to do so agam.

We support H.R. 2699 for the following reasons:

1. We need to base decisions on Nuclear Waste on the Facts.
   The State of Nevada says that the proposed repository at Yucca Mountain would be unsafe. Presumably they have studies that back up their assertions, although I am not aware of the results of such studies being published in peer reviewed journals. The overwhelming consensus of scientists and engineers that have looked at the project have found it to be safe. Included in that consensus are the top scientists at our national laboratories, the technical staff at the Nuclear Regulatory Commission (NRC), and the scientists that worked for Nye County in our authorized oversight capacity. It is time to move past empty rhetoric and to have the science behind Yucca Mountain fully vetted by independent professionals. If the repository is found to be unsafe, as the State asserts, it will end the discussion. One wonders why the State is so determined NOT to have the science fully reviewed.
2. There is local support to complete the review.
   There are seventeen counties in Nevada. Nine of them have passed resolutions calling for an objective review of the science through the statutorily mandated licensing proceeding. Yucca Mountain is entirely within the boundaries of Nye County. We are the site county. We have the most to lose if the repository is found to be unsafe. And if it is found unsafe, we will be at the forefront to oppose building the (Honorable Paul Tonko & Honorable John Shimkus) (June 12, 2019) repository. The only way to settle the differences of opinion is to finish the license proceeding. While there is not universal consent in Nevada, there is from the local communities.
3. Calling for universal consent is an excuse to do nothing.
   To believe that any controversial project in the 21st century will get consent from every level of government - state, counties, cities, tribes - is unrealistic. To believe that the consent will stand over the years of reviews and licensing is even more unrealistic. What if a State consents but one of its Senators opposes the project? What if one Member of Congress from that State objects? A very relevant example can be found in the State of Nevada itself. In the 1970s, the Federal government was seeking interim storage sites because the repository program at that time had just failed. In 1975, in response to the Environmental Statement for the Retrievable Surface Storage Facility, the Nevada legislature passed Joint Resolution 15, which strongly supported the project and made a strong bid to have the

facility built at the Nevada Test Site. Letters supporting the Retrievable Surface Storage Facility were also submitted by the City of Las Vegas, Clark County, Nye County, and Lincoln County.

Historically, New Mexico has been receptive to nuclear waste facilities, as evidenced by the success of the Waste Isolation Pilot Plant. Very recently the new Governor of New Mexico sent a letter to the U.S. Department of Energy and the NRC opposing an interim site, throwing years of work into jeopardy. Clearly, "consensus" can change with time.

To pass legislation requiring universal consent for a nuclear waste repository before proceeding simply means nothing will happen. Such a bill should be called the "Leaving Nuclear Waste Scattered Around the Country Act."

4. Authorizing Interim Storage without a permanent repository is not a solution.

   Given the unlikelihood of getting consent for a permanent repository, if an interim site is authorized, it will become a *de facto* final resting place for nuclear waste. As a result, States will likely be unwilling to accept interim storage. This also means that any waste stored at an interim site will not be stored with the safety features and assurances that a deep geological repository would guarantee.

5. The Nuclear Waste Policy Act is the law.

   Congress has never repealed the NWPA. The act was a careful balancing act protecting the State of Nevada and the local counties while moving ahead on a national security mission. The provisions of the Act were violated by the last administration when it ended the license proceeding without proper cause. A federal court of appeals subsequently declared the action improper. H.R. 2699 makes important changes to see that the full intent of the NWPA is maintained. We support that bill. We call on Congress to fund the license proceeding and not be diverted by proposals like universal consent and interim storage without the backdrop of a permanent repository. Let's let the science prevail.

Sincerely,

Leo Blundo
Nye Country Board of Commissioners
Nuclear Water Issues Representative

* * *

May 16, 2019

Dear Speaker Pelosi and Leaders McConnell, McCarthy, and Schumer:

We are a broad coalition of labor unions, state public service commissioners, local government officials, clean energy organizations, and energy trade associations, and we urge immediate congressional action to revitalize the federal used nuclear fuel program.

Another year without progress on the Yucca Mountain repository license application and consolidated interim storage is untenable. It is time for the federal government to meet its statutory and contractual obligations. Utilities and their electricity customers have done their part. They have paid more than $41 billion into the Nuclear Waste Fund. Meanwhile, taxpayers have been saddled with the federal government's inaction, with more than $7 billion in damages having already been paid from the Judgment Fund and billions more in liability continuing to mount.

The nuclear energy industry is a powerful engine for job creation. America's nuclear power plants directly employ nearly 100,000 people in high-quality, long-term jobs. This number climbs to 475,000 when you include secondary jobs. Yet the lack of a strong used fuel management program has stymied public support of nuclear power for the existing fleet as well as advanced reactors currently under development. This is unfortunate as the overwhelming scientific evidence clearly demonstrates that used nuclear fuel and defense nuclear waste can be stored, transported, and disposed of safely.

Now is the time to end the stalemate, and for the House and Senate to work in a bipartisan manner to place the federal government on a path to fulfill its responsibilities and to unburden taxpayers of the ever-mounting liability by establishing a durable program for managing used nuclear fuel and legacy defense waste.

Sincerely,
American Nuclear Society
American Public Power Association
ClearPath Action
Decommissioning Plant Coalition
Edison Electric Institute
Energy Communities Alliance

International Brotherhood of Boilermakers, Iron Ship Builders, Blacksmiths, Forgers, and Helpers

International Brotherhood of Electrical Workers

National Association of Manufacturers

National Association of Regulatory Utility Commissioners

National Rural Electric Cooperative Association

North America's Building Trades Unions

Nuclear Energy Institute

Nuclear Waste Strategy Coalition

U.S. Chamber of Commerce

United Association of Journeymen and Apprentices of the Plumbing and Pipe Fitting Industry of the United States and Canada

Testimony of Steven A. Horsford, on Behalf of the Nevada Congressional Delegation, before the Subcommittee on Environment and Climate Change, Committee on Energy and Commerce, United States House of Representatives, Hearing on "Cleaning up Communities: Ensuring Safe Storage and Disposal of Spent Nuclear Fuel," June 13, 2019

Chairman Tonko, Ranking Member Shimkus, and Members of the Subcommittee, thank you for the organizing today's hearing to discuss challenges to nuclear waste disposal and possible solutions to the problem. I appreciate the Subcommittee's recognition of the immense gravity of the issue at hand. The nation's nuclear waste stockpile continues to grow, with no immediate end or solution in sight. As the nation continues its transition towards renewable energy, it is expected that nuclear energy will continue to play a significant role in our nation's energy sector, meaning the nation's nuclear waste disposal problem will only continue to grow in magnitude and complexity.

Concerningly, the primary policy Americans depend on to facilitate nuclear waste disposal—the Nuclear Waste Policy Act of 1982 (NWPA) and its subsequent Amendments—is outdated, misguided, and misinformed. Additionally, as it was amended in 1987, policy makers removed what few Democratic principles the NWPA was founded upon, and instead decided to force the disposal of our nation's nuclear waste on Yucca Mountain, Nevada. Despite the continued objections and opposition from the Nevada public, the

Nevada Congressional Delegation, five different Nevada Governors, impacted native American tribes and communities, Nevada's tourism industry, and reams of scientific data showing the dangers associated with Yucca Mountain, some members of Congress continue to cling to this flawed policy, hesitant to accept the responsibility that comes when we recognize that our nuclear waste problem is severe, and that the nation's primary nuclear policy is seriously flawed and inadequate.

I believe all members of Congress can agree that we must act now to address this situation. However, in so doing, members of Congress must remember the many lessons we have learned since 1982 and begin to consider policy solutions that are scientifically backed and consent-based. I look forward to working with members of this Committee to find a solution to our nation's nuclear waste problem that does not involve forcing our nation's nuclear waste on Yucca Mountain, Nevada.

Unfortunately, one of the policies to be discussed today, H.R.2699, the Nuclear Waste Policy Amendments Act of 2019, introduced by Representatives Shimkus and McNerney, fails to recognize this reality and continues to promote outdated policies and undemocratic principles. This bill would restart and expedite the licensing process for Yucca Mountain and increase the amount of nuclear waste that can be stored at a nuclear waste repository from 70,000 metric tons to 110,000 metric tons, paving the way to force all the nation's nuclear waste on Yucca Mountain. The bill would also amend the Nuclear Waste Policy Act to allow for the interim storage of nuclear waste in Nevada. Just as concerning, H.R.2699 fails to develop a consent-based process that would provide states, local governments, and native American tribes with a meaningful voice in the decision to locate a nuclear waste repository. This bill is an undemocratic and misguided attempt to force nuclear waste on Nevada; it completely disregards the lessons we have learned since 1982, dissolves the few Democratic protections provided to Nevada in the NWPA, ignores recommendations made by the Blue-Ribbon Commission established to study nuclear waste disposal, and to my greatest concern, places American citizens in danger.

Establishing Yucca Mountain as the nation's only nuclear waste repository would endanger the lives of American people, disregard Nevada's opposition to Yucca Mountain, and ignore science related to Yucca Mountain. It would pave the way to disaster and create another problem, rather than provide the solution that our nation desperately needs. To start, transporting all the nation's nuclear waste to Yucca Mountain would require the shipment of more than 110,000 metric tons of high-level radioactive material over the next fifty years, traversing more than 44 states and placing the American people at significant risk of radioactive exposure caused by possible accidents. Yucca Mountain is also located in a moderate to severe earthquake hazard zone, as defined by the U.S. Geological Survey, meaning an earthquake could destroy the potential nuclear waste repository surface facilities, releasing radioactive material that could contaminate Nevada's environment. Earthquakes could also affect the groundwater flow from the repository and increase the rate and the extent of radioactive contamination of the groundwater in Amargosa Valley, impacting the livelihood of Nevadans and Californians living near the region. In 1996, a 5.6 magnitude earthquake damaged the Yucca Mountain project's field operations center. There is no reason to assume this will not happen again. As we have learned in the past, it is unwise to assume that at some point, accidents will not happen. Largescale disasters, including the Deepwater Horizon oil spill, remind us that accidents do occur, and that when we ignore warning signs, we welcome disaster. H.R.2699 would endanger more than one million people who live within 100 miles of Yucca Mountain. The monumental impact of a disaster associated with Yucca Mountain should not be understated and it certainly should not be welcomed with open arms.

It is now well-understood that the NWPA, and its associated Amendments, created an inadequate nuclear waste program that needs considerable reworking or complete replacement. In recognition that our nation's nuclear waste policy needed reassessment, President Obama appointed a Blue-Ribbon Commission (BRC) on America's Nuclear Future to study the nation's nuclear waste problems, study the inadequacies of our nuclear waste program, and recommend solutions to the problem. In 2012, the Commission submitted its final report to the Obama Administration which summarized the Commission's findings and recommended the U.S. overhaul its nuclear waste management system. The new nuclear waste strategy recommended by the commission was founded on several key principles: 1) develop a new, consent-based approach to siting nuclear waste management facilities, 2) create a new organization dedicated solely to implementing the nuclear waste management program, and 3) construct multiple geologic repositories on a timely basis.

BRC recommendations were made in recognition of the significant opposition to Yucca Mountain from the state of Nevada and other stakeholders, and the many failures of the Department of Energy to properly and competently manage the nation's nuclear waste program. The Commission's consent-based recommendation is two-fold in nature. Firstly, it recognizes the importance of consultation with impacted communities, which helps the government better understand the impacts of its proposed actions while upholding democratic principles. Secondly, it recognizes that the nation's nuclear waste disposal problem needs to be addressed in a timely manner. This has not been and will not be the case with Yucca Mountain, which will continue to face significant grassroots, legal, and political opposition. As democratically elected lawmakers, we should all recognize the importance of democratic principles and promote policy that empowers all impacted communities to have a voice in policy discussions.

That is why I sponsored H.R. 1544, the Nuclear Waste Informed Consent Act, introduced by the Nevada Congressional Delegation. H.R. 1544 would require a written consent agreement between DOE, the repository host state, affected counties, and affected Native American Tribes before beginning construction of a nuclear waste repository. Importantly, my bill would extend consent to the State of Nevada, which has for too long been denied a voice in the nuclear waste discussion. While the bill is not a complete overhaul of the nuclear waste program and is not the only policy needed to fix our nuclear waste problem, it is an important first step to improve our nuclear waste program. A companion bill, S. 649, was introduced by Senators Catherine Cortez Masto and Jacky Rosen of Nevada. H.R. 1544 and S. 649 provide a basis for amending other proposals to create a workable approach to consent-based siting for all U.S. nuclear waste storage and disposal. I hope the Committee, in its consideration of nuclear waste proposals, will consider H.R. 1544, the Nuclear Waste Informed Consent Act as a viable part of the solution. Additionally, I commend Representative Levin and Representative Matsui for introducing legislation that would begin necessary reforms to our nuclear waste program. Importantly, these proposals are based on sound reasoning and science and would not force nuclear waste on unwilling communities.

Thank you again for recognizing our nation's nuclear waste problem and organizing this hearing to discuss improvements to our nuclear waste program. It is past time for Congress to address our nation's inadequate nuclear waste program and begin developing a program that is in the best interest of all Americans. Moving forward, I would be happy to work with my colleagues to promote sound nuclear policy that does not force nuclear waste on Nevada or any other unwilling state or community. I hope we can all accept the lessons of the past few decades and work to solve our nation's nuclear waste problem in a consent-based manner, promoting the safety and security of the American people.

Testimony for the Record, the Honorable Susie Lee (NV-03), Subcommittee on Environment and Climate Change: "Cleaning up Communities: Ensuring Safe Storage and Disposal of Spent Nuclear Fuel," June 13, 2019

Thank you, Chairman Tonko and Ranking Member Shimkus, for accepting my testimony for the record.

The safe and sustainable long-term storage of nuclear waste is an issue of critical importance, and I'm pleased that the Committee is taking steps to find a solution that works for the entire country.

It should be clear to all of us by now that the 1987 law mandating Yucca Mountain as the nation's sole repository for nuclear material is an unworkable and fundamentally flawed piece of legislation. Time and time again, studies have shown that this is not a safe place for the long-term storage of nuclear waste.

For over three decades the state of Nevada has fought the forced importation of toxic material into our state. In the courts, in Congress, and at the ballot box, Nevadans have made it clear that we have no interest in becoming the nation's dumping ground for nuclear waste.

It is clear that the Nuclear Waste Policy Act Amendments is a flawed piece of legislation, and the implementation has shown that this law must be revised. Congress has the ability to repeal or revise unworkable laws, and the history of trying to implement the NWPAA makes clear that Consent must be the cornerstone of our nuclear waste policy. Without securing a buy-in from the eventual repository's state, this will only continue to be bogged down by lawsuits and legislative fights.

The delays and costs will only continue and compound so long as Congress refuses to address the fact that Nevadans of all political stripes have consistently opposed any attempt to bring nuclear waste into our state and store it at an unsafe location. At the risk of repeating myself: we don't make it, we don't use it, and we're certainly not going to take this waste.

The past 32 years have shown that without the consent of the state where a permanent repository is located, the process will only continue to suffer additional delays. If my colleagues are serious about finding a long-term solution to store this waste, they will abandon this plan to override the will of Nevadans and begin anew the process of finding a geologically-appropriate site in a state that will accept this waste within their borders.

I agree with my colleagues on both sides of the aisle that we need to find a permanent resting place for toxic nuclear waste. But no state should be compelled to accept this toxic waste from any other state without that state's consent.

Thank you again for accepting my testimony.

Subcommittee on Environment and Climate Change, Hearing on "Cleaning up Communities: Options for the Storage and Disposal of Spent Nuclear Fuel," June 13, 2019, Mr. Robert J. Halstead, Executive Director, State of Nevada, Office of the Governor, Agency for Nuclear Projects

The Honorable John Shimkus (R-IL)

1. Please provide how much federal funding from the Nuclear Waste Fund and any other federal accounts Nevada has received and expended to participate in the Yucca Mountain licensing process, including the filing of contentions and other actions in the NRC adjudicatory process.

Response: (Revised 7.23.2019)

The U.S. Department of Energy (DOE) submitted the Yucca Mountain repository license application to the U.S. Nuclear Regulatory Commission (NRC) in June 2008. The State of Nevada began full preparation to review the license application and related documents in July 2007, and has continued to participate in the proceeding, which was suspended by the NRC in 2011, and restarted by Federal Court order in 2013 with limited funding. Congress has appropriated no new funds to DOE or NRC for Yucca Mountain licensing activities since federal Fiscal Year 2011. Nuclear Waste Fund appropriations to DOE for most prior fiscal years included funds disbursed to the State of Nevada to participate in licensing activities pursuant to the Nuclear Waste Policy Act, those funds to remain available until expended. According to DOE's Office of Civilian Radioactive Waste Management, DOE provided $97,616,609 to the State of Nevada between federal Fiscal Years 1983 and 2010 inclusive.[19]

---

[19] OCRWM, Office of Business Management, *Summary of Program Financial and Budget Information as of January 31, 2010*, page 14. http://www.state.nv.us/nucwaste/news 2018/pdf/ocrwm-budget-summary.pdf.

Between July 2007 and May 2019, the State of Nevada expended $17,920,993 from the Nuclear Waste Fund and $26,379,213 from State funds, for a total of $44,300,206, to participate in the licensing process, including filing of contentions, participation in the adjudicatory process before the Construction Authorization Boards, response to orders of and actions by the Commission, review of the NRC Staff Safety Evaluation Report, review of the NRC Staff Environmental Impact Statement Supplement on Groundwater Impacts, review of NRC Licensing Support Network documents, and participation in NRC meetings in Nevada and Maryland. No federal funds have been expended for litigation. The State of Nevada files annual certifications with DOE assuring federal funds have not been used for litigation or other prohibited activities. The State of Nevada has received no funding from the Nuclear Waste Fund since federal Fiscal Year 2010, when $4,998,720 was disbursed to the Office of Attorney General[20] to remain available until expended. [Public Law 111-84-October 28, 2009, 123 STAT. 2702; Public Law 111-85-October 28, 2009, 123 STAT. 2864] We are currently in the process of closing out our expenditures for the Nevada Fiscal Year ending June 30, 2019. Our preliminary estimate is that less than $100,000 dollars remain to be spent from the Fiscal Year 2010 disbursement of federal funds.

State of Nevada Expenditures for Participation in Yucca mountain Licensing Proceeding, Agency for Nuclear Projects and Office of the Attorney General Combined

State Fiscal Year Ending June 30	State of Nevada Funds (Dollars)	Federal Nuclear Waste Fund (Dollars)
2008	2,900,367	4,021,617
2009	1,649,442	5,163,027
2010	2,397,949	2,498,239
2011	2,228,359	952,712
2012	1,108,690	1,686,143
2013	1,063,093	1,180,443
2014	1,044,583	1,247,464
2015	1,772,159	739,890
2016	2,872,887	139,931

[20] Congress appropriated to DOE $98,400,000 from the Nuclear Waste Fund and $98,400,000 from the defense nuclear waste account for FY 2010, specifying "2.54 percent shall be provided to the Office of the Attorney General of the State of Nevada solely for expenditures, other than salaries and expenses of State employees, to conduct scientific oversight responsibilities and participate in licensing activities pursuant to the NWPA."

State Fiscal Year Ending June 30	State of Nevada Funds (Dollars)	Federal Nuclear Waste Fund (Dollars)
2017	3,355,097	0
2018	3,234,134	101,380
2019 (Preliminary)	2,752,453	190,147
Total	26,379,213	17,920,993

In addition to these Yucca Mountain expenditures, during the same period (Nevada FY 2008-2019) the Agency for Nuclear Projects received $1,364,316 through grants from the Western Governors Association (WGA) to support activities by other State of Nevada agencies involved in planning, training, and exercises to prepare for DOE shipments of transuranic waste to the Waste Isolation Pilot Plant. DOE funds WGA for the specific purpose of distributing funds to states affected by these shipments. These payments could be viewed as payments to the State of Nevada from other federal accounts, but these funds are not used for Yucca Mountain licensing activities.

Subcommittee on Environment and Climate Change, Hearing on "Cleaning up Communities: Options for the Storage and Disposal of Spent Nuclear Fuel," June 13, 2019, Mr. Lake Barrett, Former Acting Director, Office of Civilian Radioactive Waste Management, U.S. Department of Energy

The Honorable David B. McKinley (R-WV)

I would like to ask about the transportation of spent fuel and radioactive waste. Can you speak to the Department of Energy's experience with Transportation?

RESPONSE:

a. How safe are the casks for shipment?

RESPONSE: Very safe.

Spent fuel and high radioactivity wastes are transported under very strict international and United States Nuclear Regulatory Commission safety and security standards. These very robust transportation casks are designed to keep the shipments safe under extreme accident conditions involving collisions (drops and punctures), fires, and submergence in water. They are generally made of very strong steel layers from 5 to 12 inches thick. They are independently analyzed by multiple organizations with confirmatory component scale testing performed as necessary. In the United Kingdom a full-scale cask was tested to confirm its performance by being struck by a train locomotive traveling at 100 miles per hour. Similar confirmatory safety testing was previously performed here in the USA with a locomotive striking an outdated cask at a 75 miles per hour speed and a cask on a truck being driven into a bridge abutment type structure at 80 miles per hour.

These are some of the strongest containers every constructed.

The US Nuclear Regulatory Commission is responsible for transportation safety standards and they constantly review safety information and periodically produce risk reports based on the most recent information available. For example, NUREG 2125 (https://www.nrc.gov/docs/ML1403/ML14031A323.pdf) concluded in 2014 that there is very low risk from spent fuel transportation during all modes of operation, including severe accidents.

The National Academies of Sciences and Engineering National Research Council performed an in-depth review of spent fuel transportation in a 2006 report entitled, "Going the Distance-The Safe Transport of Spent Nuclear Fuel and High-Level Radioactive Waste in the United States." This comprehensive report confirmed that the risks are very low and acceptable.

b. What is the Department's experience with shipment?

RESPONSE: Excellent.

The DOE and its predecessors, over the last 50 years, have safely transported over three thousand spent fuel shipments in the USA without an accident that released any radioactivity harmful to the public or environment. The USA, as all other countries, utilize the same international safety standards and globally there have been over 25,000 successful safe spent fuel shipments.

c. My understanding, to take one example, is that the Department has overseen 12,000 shipments of radiological waste to the WIPP facility in New Mexico. What is the safety record of that?

RESPONSE: Excellent.

The Waste Isolation Pilot Plant transuranic waste disposal site has safely transported by truck more than 12,500 shipments which traveled approximately 15 million miles without any significant problems.

And what is the safety record of the shipment of spent navy nuclear fuel?

RESPONSE: Excellent.

Over the past 50 years the Naval Nuclear Propulsion Program has successfully transported over 850 spent fuel cask shipments by rail to Idaho which have traveled approximately 2 million miles safely,

The Honorable Markwayne Mullin (R-OK)

As our country focuses more on carbon neutral energy, which nuclear power is the leader in, what does the uncertainty of a finalized disposal location mean for new innovations like advanced nuclear reactors?

RESPONSE: All nuclear energy facilities, even advanced ones, will produce either spent nuclear fuel or some high-level wastes from fuel recycling (if implemented). Therefore, a disposal facility, such as a Yucca Mountain, will eventually be necessary to support either advanced or current reactor types. Having a realistic disposal program capable of safely handling and disposing of all current and future nuclear wastes is an important public trust and confidence factor to support the development and siting of these new advanced nuclear systems to provide the clean air, carbon-free electrical energy that our nation needs.

a. Navy has 99 nuclear reactors that power almost all aircraft carriers and submarines. What benefits do the military gain from using advanced nuclear reactors?

RESPONSE: The US Navy has always been a technological leader in developing advanced nuclear technologies that could be useful for Naval missions. The Naval Nuclear Propulsion Program continually maintains close contact with advancing technologies and adopts aspects as appropriate. However, in my current retired situation, I am not able to provide further information in this area.

b. What are the benefits to our national security for having a permanent storage facility, like Yucca Mountain?

RESPONSE: Nuclear technologies, for the past 75 years, have always been an important aspect of our national defense programs. Former and current DOE defense production facilities and Nuclear Navy facilities have produced and continue to produce nuclear wastes that need a permanent geologic repository disposal endpoint. Although all these wastes are being stored safely in existing facilities, these current facilities were never planned to be indefinite storage sites. These materials need to be placed, in a timely manner, into a permanent disposal facility, such as Yucca Mountain, to allow current defense related facilities to perform their current important national security or cleanup missions without interruption.

A current example is the Naval Reactors Propulsion Program's Naval Reactor Facility at the DOE Idaho National Engineering Laboratory (INEL). This facility receives all spent nuclear fuel from naval reactors for research and development purposes and final preparation for disposal in a permanent disposal geologic repository. Due to the concerns of the State of Idaho about indefinite storage of used or spent nuclear fuels in Idaho, the DOE and Naval Reactors Propulsion Program signed a settlement agreement in 1995 with the State of Idaho that would allow continued limited Naval fuel shipments to Idaho and that current stored spent fuel there would be removed before 2035. The only potential repository site available to remove this Idaho fuel, within this agreement time frame, would be Yucca Mountain.

Compliance with this agreement is important to continue the successful long-standing history for Idaho to receive nuclear navy spent fuel and any significant interruptions in these fuel receipts could impact fleet readiness, thus becoming a national security concern.

c. How is the military's nuclear waste currently transported to its temporary storage facility in Idaho?

RESPONSE: Navy spent nuclear fuels are transported to Idaho by rail in specially designed transportation casks.

d. Has there ever been any issue transporting this waste?

RESPONSE: There has never been any public health or safety problems with the shipments. Over the past 60 years, there have been over 850 shipments that have safely traveled approximately 2 million miles from east coast and west coast naval facilities to Idaho.

There has been an Idaho social/political concern that DOE has not been able to meet the expectations of the State of Idaho to remove stored navy and commercial spent fuels from the INEL after research and development activities have been completed. This general concern had manifested into some jurisdictional concerns with nuclear transportation access into the INEL site. These have all been successfully resolved with the 1995 agreement and other agreements. However, if there are significant future DOE delays in meeting the 1995 agreement to remove spent fuels from Idaho by 2035, such complications may possibly arise in the future.

e. Why is there an issue transporting this waste from its temporary waste in Idaho to a permanent site in Nevada?

RESPONSE: There is no transportation problem preventing the movement of Idaho fuels or wastes to a geologic repository site, other than there is no repository site to go to.

The problem is that Congress is not funding the DOE and NRC to allow completion of the NRC licensing process that is necessary to be able to start construction of the Yucca Mountain permanent repository. If the DOE 2008 Yucca Mountain repository site construction authorization application is approved by the NRC for construction, and Congress supports construction with funding, then transportation details can be successfully developed under existing laws, regulations and procedures to allow transportation from Idaho to Yucca Mountain.

If Yucca Mountain funding is reestablished soon, fuel could be removed from Idaho by the 2035 agreement date thus resolving this national security concern.

The Honorable Bill Johnson (R-OH)

Mr. Barrett, you expressed confidence in the scientific and technical merits of the Yucca application. Politically, it is challenging. Many of the actions that led to Yucca took place in the late 1970s and 80s. There was considerable work performed then and experience about what is necessary for developing a durable program. Can you elaborate on that experience?

RESPONSE: Our nation has studied geologic formations for many potential repository sites across the country for nuclear waste disposal for over 50 years with the best scientists available. For example, detailed studies started at Yucca Mountain in 1978 as well as at dozens of other sites and regions across the country. In the mid-1980s the DOE studied in detail nine sites in five states and published thousands of pages of scientific information in draft form for public comment before nominating five sites and finally recommending three sites (Yucca Mountain NV, Hanford WA, and Deaf Smith TX) for detailed site characterization. These were all scientifically "good" sites, with each of course having greater and lesser attributes. However, Congress in 1987, after reviewing the scientific work on all the sites, stated by statute, that only the Yucca Mountain site would be further studied.

From 1988 until the 2002 Presidential Site Recommendation, the DOE team, following the law, performed extensive scientific exploration and testing deep inside of Yucca Mountain to analyze the technical aspects of the site and its repository design to be able to demonstrate that it could meet the very protective EPA and NRC regulatory safety and environmental protection standards. The conclusion of the DOE science team, as confirmed by international peer review, was that the site was acceptable, and it was recommended to Congress for designation under law by President Bush.

From 2002 until 2008 the DOE prepared an extensive formal license application to submit to the Nuclear Regulatory Commission that provided much more scientific justification that the site would meet all safety and environmental protection requirements.

Even after the DOE submitted the Yucca Mountain license application in 2008, confirmatory scientific work continued that has indicated that there was considerable conservatism in the DOE work, meaning that the site would perform even better than predicted. For example, a detailed US Geologic Survey report on earthquake risks in the Yucca mountain area (https://pubs.usgs.gov/of/2013/1245/pdf/of2013-1245.pdf) confirmed that Yucca Mountain was well protected from any potential earthquake that could happen in the area or beyond. In fact, the recent July 2019 California earthquakes that that were felt in Nevada were not unexpected and would likely have no impact on a Yucca Mountain repository. Based on all the scientific studies, Yucca Mountain would meet all seismic public protection standards for any long-term projected earthquake event.

From a scientific earthquake protection point of view, a passive deep underground repository is one of the lowest risk places if there is a large earthquake. Even assuming if some magical scientifically unsupported earthquake were to happen in the future, an underground Yucca Mountain repository would be much less impacted than other structures in the Las Vegas area. For example, if such an unscientific earthquake occurred, most structures in Las Vegas, including most high-rise buildings and residential homes, would be far more likely to sustain major structural damage than would a repository at Yucca Mountain, which would likely survive with minimal impact.

From 2008 to 2012, when the Yucca program was stopped, there was an intensive independent Nuclear Regulatory Commission staff safety review of the DOE Yucca Mountain license application. The conclusion of this detailed scientific review was that Yucca Mountain met the regulatory standards necessary to ensure public health and safety and environmental protection.

All the above extensive scientific work is why I have confidence in the scientific merits of the Yucca Mountain site and why I believe the final stage of the licensing process should be finished.

It is my view that if the nation were to technically develop a new different site somewhere else in this country, that the technical effort would be similar if not greater. New standards would be required, and detailed underground work would be just as complex, if not more so.

If there is a future repository siting program, more attention will likely be needed to maintain social/political support, especially in the area of federal-state level relationships. It is difficult to project, but these will likely be a substantial added cost which will be in addition to extensive technical-scientific safety and environmental protection work that is always necessary.

I question how we could possibly do any type of siting work if we started over and began again today. Mr. Barrett, in this age of 24/7 social media, aren't there a whole new set of challenges to ensure people have full and accurate information about siting and technical determination of repository.

RESPONSE: Yes, it is much more challenging these days to communicate useful information to the public in a meaningful way when there is so much willful interference by those with strong disruptive social-political agendas being so active in today's many media platforms. Modern internet and social media communication modes allow special interests to quickly spread inaccurate and misleading sensational negative information that erodes public confidence in any complex governmental or establishment program they wish to attack. It is much more difficult today for responsible implementing organizations to effectively communicate complex safety and environmental protection scientific information when detractors can post distorted and often false information without any repercussions. Although the implementing organization can now promptly put all their factual scientific information online, it is very complex and not easily understood by the general public. Detractors can easily undermine good products by posting inaccurate partial truth "sound bites" that have little basis but can stir strong emotions that are very detrimental to the public trust and confidence in implementing organizations.

a. Would information challenges like this be addressed by completing the NRC licensing process?

RESPONSE: Yes. So much has already been done and everything is already in the public record with Yucca Mountain. It is time to complete the licensing process to conclude, through independent judges, if the site meets the regulatory requirements or not. And if it does, as I expect, then move on to the next national decision point of Congress deciding whether to provide the funding necessary to build the repository to receive spent fuel from the reactor and waste sites spread across the country there or not.

The issues with Yucca Mountain are well known and well-studied, so now is the time to just finish the legal process and then politically decide what the next steps should be. Starting over will take many decades and will cost many tens of billions of dollars in developing a new site and paying for continued storage at stranded sites across the nation. This challenge does not get easier with time.

b. Does it make any sense to stop short now?

RESPONSE: No. As a nation, we need a solution to our fast-growing nuclear waste challenge. There is no better realistic option that can replace Yucca Mountain in a reasonable timeframe. Other options, such as trying to develop a different consent-based repository site, can be added in parallel, which I support, but that should not just replace Yucca Mountain.

c. What would be lost to the public, ratepayers, taxpayers if we turn away from the $15 billion-dollar investment in Yucca without completing the licensing?

RESPONSE: All the Yucca Mountain site work value would be lost and the costs to our taxpayers will increase by tens of billions of dollars more to maintain spent fuel and wastes at isolated storage sites spread across the country on our seacoasts, rivers, and lakes for our lifetimes or more. We would be consigning a major debt and significant burden upon our grandchildren and great grandchildren for no real reason.

Another factor is that it could be very difficult to attract future scientists to work on such a politically sensitive program as nuclear fuel and high-level radioactive waste disposal in the future. Thousands of excellent scientists have worked very hard on the Yucca Mountain program and have been unfairly maligned from a social/political perspective. Seeing good scientific work dismissed by often emotional political outbreaks is difficult to endure and, in the future, will make it challenging to assemble a new team to achieve what the nation needs to have done.

Congress should stand up to allow the final judgment process for Yucca Mountain to finish in respect for the scientists that worked very hard for decades. It is demoralizing to work so hard to see everything lost, without scientific reason, due to relatively short-term political driven Not-in-My-Backyard emotions.

d. What would that mean for future efforts to site a repository?

RESPONSE: In my view, if Yucca Mountain cannot be completed, I doubt that any other site in the nation can be realistically developed due to state level social/political resistance. Yucca is an excellent isolated technical site that has been well studied for over four decades. There may be other scientifically potentially good geologic sites across the nation, however there is no reason to believe that the same state level Not-In-My Backyard social/political resistance would not be met there as well.

The 1982 Nuclear Waste Policy Act established a statutory process for the balance between state level rights and federal level needs for establishing a necessary repository somewhere in the nation. This balance required DOE to fund potential state level involvement and also allowed the potential host state to statutorily "disapprove" the site, for any reason, and that the site was to be abandoned unless both US Senate and House of Representatives voted to override the state disapproval. In 2002, Nevada disapproved the Yucca Mountain site and Congress overrode the disapproval, however the State of Nevada has never agreed and has been able to Congressionally prevent funding to complete the Yucca Mountain process, thus blocking a national need solution. This has happened even though the local governments, nearest the Yucca site, support continuing work on the project. I do not see how this would be different with any other potential host state in this country in the future.

Chapter 2

High-Risk Radioactive Material: Opportunities Exist to Improve the Security of Sources No Longer in Use*

United States Government Accountability Office

Abbreviations

CIRP	Cesium Irradiator Replacement Project
CRCPD	Conference on Radiation Control Program Directors, Inc.
DOE	Department of Energy
DSWG	Disused Sources Working Group
GTCC	Greater Than Class C
IAEA	International Atomic Energy Agency
NNSA	National Nuclear Security Administration
NRC	U.S. Nuclear Regulatory Commission
NSTS	National Source Tracking System
OSRP	Off-Site Source Recovery Program
WIPP	Waste Isolation Pilot Plant
WINS	World Institute for Nuclear Security

* This is an edited, reformatted and augmented version of the United States Government Accountability Office Report to the Committee on Armed Services, House of Representatives, Publication No. GAO-24-105998, dated November 2023.

In: Nuclear Waste
Editor: Ken Ykstra
ISBN: 979-8-89113-719-6

Why GAO Did This Study

Radioactive sources are commonly used for medical, industrial, and research purposes. However, these materials can be harmful and dangerous, if used improperly.

NRC and states to which it has delegated authority issue licenses for the possession and use of radioactive sources. These entities regulate disposal facilities that can accept certain sources and waste. Several federal programs support disposal of some sources, but some licensees still hold onto sources beyond their useful lives. Doing so increases the risk that sources could be orphaned and misused, for example, in a dirty bomb.

House Report 117-118 accompanying a bill for the National Defense Authorization Act for Fiscal Year 2022 includes a provision for GAO to review the disposition of radioactive sources. This report examines (1) the factors that contribute to licensees delaying disposal of disused high-risk radioactive sources, and (2) leading practices that, if implemented, could help address challenges related to the disposal of some disused radioactive sources. GAO reviewed relevant laws, regulations, and key organizations' documents on leading practices. GAO also interviewed agency and industry officials and conducted site visits.

What GAO Recommends

GAO is making three recommendations, including that NRC should assess leading practices that would minimize the time that disused sources are in a licensee's possession. NRC neither agreed nor disagreed with this recommendation, and the agencies generally agreed with the other two recommendations.

What GAO Found

Licensees of high-risk radioactive sources may delay disposing of sources that are in their possession but no longer in use (i.e., disused) for a variety of reasons. For example, the U.S. Nuclear Regulatory Commission (NRC) does not require licensees to dispose of radioactive sources unless a licensee is terminating all activities under its license at specific locations. In addition, some high-risk sources containing radioactive materials that have a long life cycle, including cesium-137 and americium-241, have limited disposal pathways that may require government assistance or may not have a viable disposal pathway at all. Specifically, sources used in the oil and gas industry that contain americium-241 of foreign origin currently have no permanent disposal pathway, leaving them vulnerable to loss or abandonment.

GAO identified leading practices supported by key entities—such as the International Atomic Energy Agency—that are not reflected in NRC requirements and could help address some disposal challenges. Thesepractices include tracking sources, imposing limits and fees on possession, or collecting financial assurances at the time a source is purchased to offset later disposal costs. Assessing adoption of these leading practices nationwide may more broadly incentivize timely disposal, potentially reduce overall cost to the government, and reduce the risk that radioactive sources could be used in a dirty bomb.

Disused High-Risk Radioactive Sources at a Source Processing Facility

Source: GAO | GAO-24-105998.

November 30, 2023

The Honorable Mike Rogers
Chairman

The Honorable Adam Smith
Ranking Member
Committee on Armed Services House of Representatives

Radioactive sources—such as americium-241, cesium-137, cobalt-60, and iridium-192—are commonly used throughout the U.S. for medical, industrial, and research purposes. Those engaged in these purposes, often referred to as licensees, are licensed by a federal or state regulator to safely and securely possess and use radioactive sources. However, these materials can be harmful and dangerous if used improperly. For example, terrorists could use even a small amount to construct a dirty bomb, which uses conventional explosives to spread radioactive material. Furthermore, some radioactive sources are used in industries subject to boom-and-bust cycles, raising the potential for them to become orphaned. According to the U.S. Nuclear Regulatory Commission (NRC), which regulates radioactive sources, orphan sources include those that are not under regulatory control and require removal to protect public health and safety from a radiological threat.[1]

In March 2023, President Biden signed a national security memorandum to establish a more comprehensive policy for securing radioactive sealed sources.[2] In addition, the memorandum calls for the permanent disposal or recycling of certain disused and unwanted radioactive sources, which could be used in a dirty bomb.

Radioactive sources typically have three phases in their life cycle: (1) the licensee is in possession of and using the source; (2) the licensee is still in possession of the source, is no longer using it, and has not yet disposed of it

[1] The International Atomic Energy Agency (IAEA) partners worldwide to promote the safe, secure, and peaceful use of nuclear technologies. According to the IAEA, an orphan source is a radioactive source that poses sufficient radiological hazard to warrant regulatory control but that is not under regulatory control because it has never been so, or because it has been abandoned, lost, misplaced, stolen, or otherwise transferred without proper authorization.

[2] The White House, "FACT SHEET: President Biden Signs National Security Memorandumto Counter Weapons of Mass Destruction Terrorism and Advance Nuclear and Radioactive Material Security" (Mar. 2, 2023).

(i.e., the source is disused);[3] and (3) the licensee hasdisposed of the source, and the source is secured at an approved disposal facility (see Figure 1).

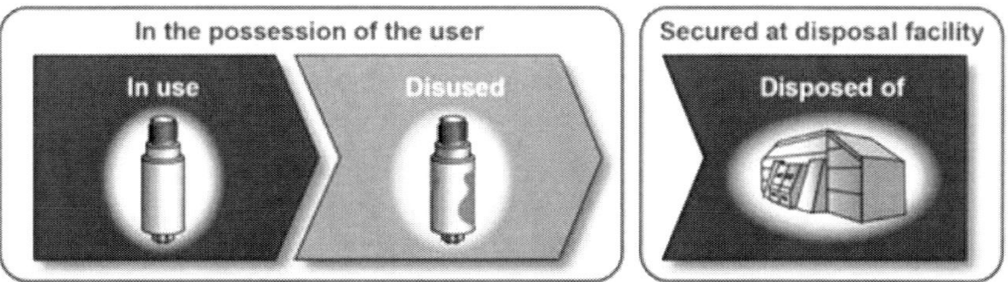

Source: GAO analysis and illustrations. | GAO-24-105998.

Figure 1. Life Cycle of Sealed Radioactive Sources.

Although radioactive sources have a defined period of economic viability based on their half-life and physical condition, licensees sometimes delay disposal of the source, leaving it in the disused phase.[4] In certain circumstances, holding onto radioactive sources in a disused state can increase the risk that sources could become orphaned and raises the potential that a source could be stolen and used in a dirty bomb. This is particularly problematic for sources such as cesium-137 and americium- 241, as these sources remain dangerous for many years.[5] Furthermore, we have previously reported that these materials, if stolen and used in a dirty bomb, could result in billions of dollars in socioeconomic damage.[6]

Federal and state entities are both involved in regulating and securing radioactive sources. NRC and certain states, known as agreement states, issue licenses to persons who need to possess and use radioactive materials, and regulate the safety and security of these sources.[7] Agreement states also issue

---

[3] For the purposes of this report, we will refer to sources that are in this second phase as"disused."

[4] Half-life is the length of time it takes for half of the radioactive atoms of a specificradionuclide to decay.

[5] The half-life of cesium-137 is 30.08 years. The half-life for americium-241 is 432.6 years.

[6] These costs could include extensive environmental cleanup, loss of access to homes and business, and deaths from evacuations. GAO, *Combating Nuclear Terrorism: NRC Needs to Take Additional Actions to Ensure the Security of High-Risk Radioactive Material*, GAO-19-468 (Washington, D.C.: Apr. 4, 2019).

[7] The Atomic Energy Act of 1954 gives the U.S. Nuclear Regulatory Commission regulatory authority over domestic industrial, medical, and research uses of radioactivematerials. The act also authorizes the NRC to enter into agreements with states (called"agreement states") so they assume, and NRC discontinues, regulatory authority over specified radioactive materials. There are currently 39 agreement states, and the remaining states are known as "NRC states." The NRC and agreement states act as regulators. They license, monitor, track, and require security for radioactive materials inorder to protect both workers and the

licenses that regulate disposal facilities that accept certain commercially generated waste.[8] The Department of Energy (DOE) is responsible for identifying a disposal pathway, or disposal option, for a certain type of radioactive waste that does not currently have a commercial disposal pathway.[9] DOE's National Nuclear Security Administration (NNSA) has multiple programs that assist private entities either with disposal of certain disused sources or with facilitating transition to alternative technologies. In addition, the NRC and NNSA support the Conference of Radiation Control Program Directors, Inc., (CRCPD) which subsidizes disposal of radioactive sources at commercial disposal facilities and picks up orphaned sources.[10]

House Report 117-118, accompanying a bill for the National Defense Authorization Act for Fiscal Year 2022, includes a provision for us to review the disposal of radioactive sources and the potential to incentivize private industry to dispose of radioactive sources.[11] Our report examines (1) the factors that contribute to licensees delaying the disposal of disused high-risk radioactive sources; and (2) leading practices that, if implemented, could help address challenges related to the disposal of some disused radioactive sources.

To evaluate the factors that contribute to licensees delaying the disposal of disused high-risk radioactive sources, we reviewed legislation, regulations, and guidance concerning storage and disposal of radioactive sources. During our initial meetings with the NRC and NNSA, we asked for information on potential contacts to interview. They provided us with approximately 20 contacts, including a wide range of agency officials, state regulators, and industry representatives from small and large licensees. Initial contacts gave us additional sources.

During the engagement, we conducted over 30 interviews to understand the challenges faced by licensees possessing different sources in the disused phase. Specifically, we interviewed officials from the NRC, DOE, NNSA, and six agreement states. We also interviewed six licensees, several radioactive

public from exposure to hazardous levels of radiation generated by the activities of licensees.

8 Existing commercial low-level waste disposal facilities are located in agreement states, including: EnergySolutions Barnwell Operations, located in Barnwell, South Carolina; U.S. Ecology, located in Richland, Washington; EnergySolutions Clive Operations, located in Clive, Utah; and Waste Control Specialists, LLC, located near Andrews, Texas.

9 Disposal pathways address the various means for safely disposing of these materials sothat they will not represent a health and safety risk to current and future populations.

10 CRCPD is a 501(c)(3) nonprofit nongovernmental professional organization dedicated to radiation protection.

11 H.R. Rep. No. 117-118, at 338 (2021) (accompanying H.R. 4350, a bill for the NationalDefense Authorization Act for Fiscal Year 2022).

source manufacturers and processors, two industry groups, and one source broker. Using the information we obtained from the interviews, we identified several challenges licensees face that can result in delays to disposing of disused radioactive sources. The findings from these interviews are not generalizable; however, they highlight some common challenges experienced by different groups of licensees and as perceived by different types of stakeholders.

We also traveled to Texas and met with individuals at small, medium, and large private companies that use radioactive sources; source processors; and a private waste disposal facility to observe how they manage sources that are no longer in use. We chose Texas because of the significant size of the oil and gas industry operating in the state and because it has one of the few private waste disposal facilities in the U.S. Finally, we reviewed data from NNSA used to estimate the number of disused radioactive sources in the U.S. We interviewed officials to gain an understanding of the estimate and determined that the data were sufficiently reliable for our purpose of estimating the number of disused foreign-origin americium- 241 sources in the U.S.

To evaluate which leading practices certain entities have implemented to help address challenges related to the disposal of some disused radioactive sources, we reviewed relevant documents to identify leading practices recommended by key entities, including the International Atomic Energy Agency (IAEA), the Disused Sources Working Group (DSWG), and the World Institute for Nuclear Security (WINS).[12] We also interviewed officials from six agreement states identified by the Disused Sources Working Group and NNSA officials, some of which have implemented leading practices that could help address challenges associated with disposal of radioactive sources.[13] We analyzed the results from these interviews to determine what leading practices have been implemented by the different agreement states that address disposal challenges. These results are not generalizable to all agreement states.

[12] The IAEA is the world's central intergovernmental forum for scientific and technical cooperation in the nuclear field. It promotes the safe, secure, and peaceful uses of nuclear science and technology. The Disused Sources Working Group is comprised of eight directors of the Low-Level Radioactive Waste Forum. The group solicits input from a broad range of stakeholders, issues reports, and makes recommendations about disposal of radioactive materials. WINS is an international organization whose mission is to improve professionalism and competence in nuclear security worldwide. WINS publishes best practices guides related to nuclear security.

[13] We spoke to officials from the following agreement states: Colorado, Florida, Illinois,New Jersey, Oregon, and Texas.

We interviewed agency officials at the NRC and NNSA, licensees possessing radioactive sources, the CRCPD, and industry representatives—as identified above—to understand the potential for implementing different mechanisms to incentivize disposal of radioactive sources. Finally, we attended the Low-Level Radioactive Waste Forum and Disused Sources Working Group meetings in October 2022 and March 2023, where we attended sessions related to leading practices for disposal of radioactive materials.

We conducted this performance audit from April 2022 to November 2023 in accordance with generally accepted government auditing standards. Those standards require that we plan and perform the audit to obtain sufficient, appropriate evidence to provide a reasonable basis for our findings and conclusions based on our audit objectives. We believe that the evidence obtained provides a reasonable basis for our findings and conclusions based on our audit objectives.

Background

Uses for Radioactive Sources

Radioactive sources are used in various industrial and medical processes. In the U.S., the radioactive sources most commonly used for medical, industrial, and research purposes are americium-241, cesium-137, cobalt-60, and iridium-192. For example, americium-241 is used in well logging to examine geologic features around a borehole or well as an aid to searching for oil, gas, and water or conducting environmental or other forms of underground monitoring. Well logging devices are lowered downhole and emit radiation to take readings on the characteristics of an underground formation, such as its chemical and mineral contents. In the medical industry, cesium-137 is widely used in blood irradiators—where donor blood is exposed to radiation, which inactivates a type of white blood cell that may fatally complicate transfusion for some recipients. The most common method of using radiation to treat blood is to place blood bags into a shielded chamber inside of an irradiator containing cesium-137. Cesium-137 is also used in research irradiators to expose cell cultures or animal specimens to gamma radiation. Research irradiators are used to study DNA damage, immune response, cancer development, and other areas. Ultimately, these radioactive sources will reach the end of their lives and need disposal.

Categories of Radioactive Sources and Types of Waste

The U.S. and other nations endorse IAEA's Code of Conduct on the Safety and Security of Radioactive Sources, which establishes basic principles and guidance for the safe and secure use of radioactive sources that are dangerous to human health. It ranks radioactive sourcesinto one of five categories based on their potential to harm human health,if not safely managed or securely protected:

- Category 1 sources, the most dangerous, are defined as having an activity at least 1,000 times or more than the activity likely to cause permanent human injury. A category 1 source would likely cause permanent injury if handled for more than a few minutes.
- Category 2 sources are defined as having an activity at least 10 times, but less than 1,000 times, the activity likely to cause permanenthuman injury to a person who handles them. A category 2 source could cause permanent injury if handled for a short time (minutes to hours).
- Category 3 sources are defined as having an activity at least the minimum amount but less than 10 times the activity sufficient to causepermanent human injury. A category 3 source could cause permanentinjury if handled for some hours.
- Category 4 and 5 sources are unlikely or unable to cause permanent human injury.

The NRC requires enhanced security controls for licensees possessing category 1 and 2 radioactive materials. This includes national tracking of certain individual radioactive sources in its National Source Tracking System (NSTS).[14] However, these security controls do not extend to categories 3, 4, and 5. As we have reported in the past, NRC has a threshold of category 1 and 2 sources as part of its requirements to track sources in NSTS.[15] As a result, NRC is not able to easily determine how many category 3, 4, and 5 sources there are in the U.S. nor easily identify where they are located.[16] We have

---

[14] NSTS is a transaction-based system that tracks each major step that category 1 and 2radioactive sources take within the U.S.

[15] GAO, *Nuclear Security: NRC Has Enhanced the Controls of Dangerous RadioactiveMaterials, but Vulnerabilities Remain*, GAO-16-330 (Washington, D.C.: July 1, 2016).

[16] According to NRC officials, the NRC and agreement states issue licenses that include a condition to indicate the location of use for the radioactive sources. However, this information for category 3, 4, and 5 sources is not centralized.

previously recommended that NRC extend its requirements for national source tracking to category 3 sources, given the risks they can pose.[17]

Radioactive sources are disposed at certain facilities based on what types of waste those facilities are set up to accept. NRC regulations classify

low-level radioactive waste as Classes A, B, C, or Greater Than Class C (GTCC).[18]

- Class A has the lowest radiological hazard and can contain short-lived radionuclides that decay to background levels within a few decades.
- Class B contains higher concentrations of short-lived radionuclides than Class A.
- Class C contains higher concentrations of both short-lived and long-lived radionuclides than Class B low-level radioactive waste.
- GTCC has concentrations of certain radionuclides that exceed the Class C limits. GTCC is the most radiologically hazardous within the low-level radioactive waste classification.

Waste Disposal Pathways

Although the NRC considers all four of the most commonly used radioactive sources for medical, industrial and research purposes— americium-241, cesium-137, cobalt-60, and iridium-192—to be high-risk radionuclides, different disposal pathways are available based on the concentration and half-life of radionuclides when these sources becomedisused. Typically, the waste from cobalt-60 and iridium-192 sealed sources can be disposed of at commercial waste facilities, though licensees often return them to suppliers through routine source exchanges that maintain the continued functionality of the device.

By contrast, many cesium-137 and americium-241 sources are more challenging to dispose. Specifically, the half-lives of cesium-137 and americium-241 are much longer than those of cobalt-60 and iridium-192, and

---

[17] GAO-16-330. This recommendation has not been implemented.

[18] 10 C.F.R. §§ 61.55(a)(2), 72.3. There is no specific crosswalk between the differentcategories of radioactive sources and the different classes of radioactive waste. Categories of radioactive sources are based on the relative radioactivity of the materials and their impacts on human health. Classes of waste are based on the concentration andhalf-life of the material. Furthermore, DOE does not use the NRC's waste classification system and instead has different waste acceptance criteria for each of its disposal facilities.

the concentrations that would qualify as Class A, B, or C waste are likely much lower than quantities typically used in sealed sources. Thus, the quantity of material in many of the larger sealed sources used in the medical and oil and gas industries would likely render them GTCC waste.Figure 2 shows the different disposal pathways for these radioactive sources.

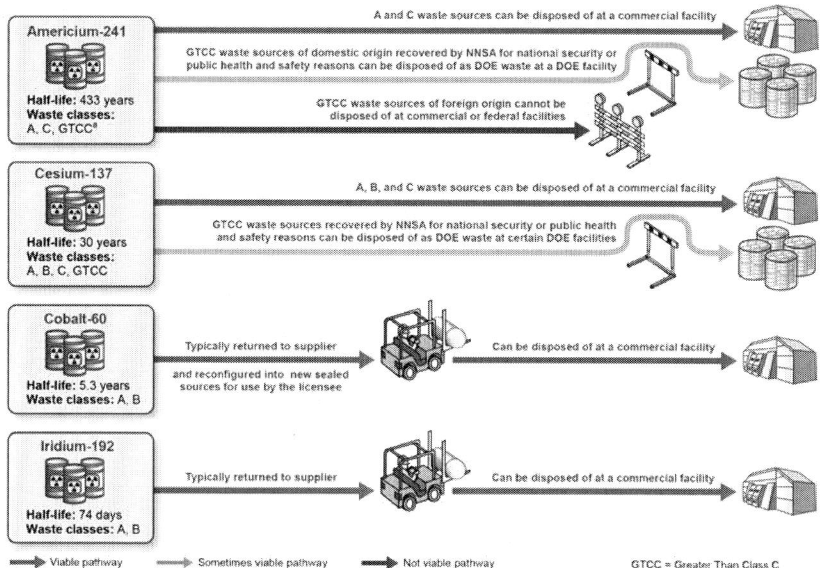

Source: GAO analysis and illustration of DOE and industry information. | GAO-24-105998.

Note: GTCC waste cannot be disposed of at a federal facility unless its ownership is first transferred to the federal government and it meets the waste acceptance criteria for disposal at that facility.

When disused sealed sources are recovered by the National Nuclear Security Administration (NNSA) for national security or public health and safety reasons, transfer of ownership from the source ownerto the Department of Energy (DOE) is officially documented through an Authorization to Transfer/Relinquishment of Ownership/Custody form. These sources may be disposed of at DOE facilities as low-level radioactive waste or transuranic waste if the sources meet the waste acceptance criteria for those facilities.

[a] There is no Class B limit for americium-241 under 10 C.F.R. § 61.55.

Figure 2. Disposal Pathways for Americium-241, Cesium-137, Cobalt-60, and Iridium-192.

Currently, there are commercial facilities in the U.S. that are licensed by agreement states to dispose of Class A, B, and C waste, but there is no commercial facility that can generally accept GTCC waste.[19] Instead, GTCC waste must generally be disposed of in a geologic repository, unless the NRC approves an alternative disposal facility on a case-by- case basis.[20] The nation currently has one geologic repository, the Waste Isolation Pilot Plant (WIPP), in Carlsbad, New Mexico. However, by law, WIPP can only accept transuranic radioactive waste generated by atomic energy defense activities.[21]

In some cases, the federal government can take GTCC from commercial entities and dispose of it at a federal facility. When GTCC is recovered by NNSA for national security or public health and safety reasons, ownership is transferred from the source owner to DOE. Waste owned by DOE may be disposed of at a federal facility as low-level radioactive waste or transuranic waste if it meets the waste acceptance criteria for that facility. Not all such waste may meet waste acceptance criteria, however.

Specifically, foreign-origin americium-241 may not be suitable for near surface disposal and also may not meet the waste acceptance criteria for WIPP. Under certain circumstances, waste from materials used commercially may qualify as defense waste, but DOE has determined that foreign-origin sources cannot be categorized as defense waste and cannot be disposed of at WIPP. Thus, in certain circumstances, costs for low-level waste that cannot be

---

[19] The federal government is responsible for GTCC disposal under the Low-Level Radioactive Waste Policy Amendments Act of 1985 (1985 Act). Pub. L. No. 99-240 § 102,99 Stat. 1842, 1844 (1986) (codified at 42 U.S.C. § 2021c(b)(1)(D)). We reported in 2022 that DOE is required by statute to await congressional direction before proceeding with a decision on GTCC disposal for nondefense waste. We suggested that Congress provide direction to DOE on nondefense GTCC waste disposal so that DOE can proceed with adecision. GAO, *Nuclear Waste: DOE Needs to Improve Transparency in Planning for Disposal of Certain Low-Level Waste*, GAO-22-105636 (Washington, D.C.: Sept. 29, 2022). As of September 2023, Congress has not acted on this matter.

[20] The NRC defines a "geological repository" as an excavated underground facility designed, constructed, and operated for safe and secure permanent disposal of high-levelradioactive waste. The NRC has taken steps to update its rules regarding GTCC waste disposal. In 2015, the Commission directed NRC staff to prepare a regulatory basis for the disposal of GTCC waste through means other than deep geologic disposal, including near-surface disposal. The NRC issued a draft regulatory basis in 2019 stating that approximately 80 percent of the total volume of all GTCC waste is potentially suitable fornear-surface disposal, as long as appropriate controls are implemented and a sufficient site-specific analysis is conducted. In April 2022, the Commission voted to proceed with rulemaking and develop guidance specifically for the near-surface disposal of GTCC waste.

[21] The Atomic Energy Act defines transuranic waste as "material contaminated with elements that have an atomic number greater than 92, including neptunium, plutonium,americium, and curium, and that are in concentrations greater than 10 nanocuries per gram, or in such other concentrations as the [NRC] may prescribe to protect the public health and safety."

commercially disposed of, such as GTCCwaste from domestic americium-241 and larger cesium-137 sources, areborne by the government through NNSA activities.[22]

NNSA has one program, the Off-Site Source Recovery Program (OSRP), that removes disused sources from licensees and disposes of the material at federal facilities. OSRP typically removes large high-risk sources from licensees to protect against the material being stolen and used in a dirty bomb. NNSA also has the Cesium Irradiator Replacement Project (CIRP), which provides an incentive to remove and replace cesium-137 irradiators from medical and industrial facilities with non- radioisotopic x-ray devices. NNSA also sponsors state-level programs runby CRCPD to provide a subsidy for the disposal of disused sources with commercial disposal options and to help address the recovery of orphan sources.[23]

Risks Posed by Radioactive Sources Outside of Regulatory Control

As we have reported, category 1, 2, and 3 sources could produce billionsof dollars of socioeconomic damage if released through a dirty bomb, though the NRC's enhanced security requirements only cover category 1and 2 sources.[24] Furthermore, our work has demonstrated that the risks of a dirty bomb are increasing. As we reported, the risk of a dirty bomb isdetermined by the function of three components: threat, vulnerability, andconsequence. Threat is generally defined as entities or actions with the potential to cause harm—including terrorist attacks. We previously reported that NNSA assessments of the threat environment show an increasing interest in using radioactive material for making a dirty bomb.[25]

The second component of risk from a dirty bomb, vulnerability, includes physical features or operational attributes that render an asset open to

[22] DOE classifies radioactive waste as low-level, high-level, or transuranic waste.

[23] CRCPD runs the Source Collection and Threat Reduction Program, which picks upsmaller radioactive sources for disposal.

[24] GAO-19-468. We made three recommendations to the NRC, including that it consider socioeconomic consequences and fatalities from evacuations as criteria for determining security measures and require additional security measures for high-risk quantities of certain category 3 radioactive materials. The NRC generally disagreed with the recommendations. However, we continue to believe these recommendations are important.

[25] GAO, Alternatives to Radioactive Materials: A National Strategy to Support Alternative Technologies May Reduce Risks of a Dirty Bomb, GAO-22-104113 (Washington, D.C.: Oct. 21, 2021).

exploitation, including gaps in security measures such as gates, locks, perimeter fences, and computer networks. Our work shows the persistence of vulnerabilities in the security of radioactive materials, particularly for category 3 sources that are not subject to the NRC's enhanced security requirements.[26] Moreover, the potential for orphan sources could create additional vulnerabilities. Finally, the third component of risk from a dirty bomb, consequence, includes the effectsfrom terrorist attacks that could result in impacts to public health and safety and the economy.

Various Factors Contribute to Delays in Disposal of Disused Sources

We identified various factors that contribute to delays in disposal of certain disused sources. First, the NRC does not require licensees to dispose of radioactive sources unless a licensee is terminating all activities under its license at specific locations. Second, licensees do not have an option to dispose of GTCC from foreign-origin americium-241 sources. Third, licensees may delay disposal of GTCC from cesium-137 sources and may rely on government assistance because of the high costof disposal and the absence of a GTCC disposal facility in the U.S.

NRC Does Not Require Licensees to Dispose of All Disused Sources but Has Taken a Step to Promote Disposal

The NRC's regulations cover a number of safety and security requirements for licensees in possession of radioactive sources. However, the NRC's

[26] GAO, *Preventing a Dirty Bomb: Vulnerabilities Persist in NRC's Controls for Purchasesof High-Risk Radioactive Materials*, GAO-22-103441 (Washington, D.C.: July 14, 2022). We made two recommendations to enhance the security of category 3 sources. We recommended that the NRC (1) immediately require vendors to verify category 3 licenses with the appropriate regulatory authority and (2) add security features to its licensing process that improve the integrity of the process and make it less vulnerable to altering or forging licenses. To address our recommendations, the NRC said it would propose a rulemaking to strengthen licensing. However, vulnerabilities will remain until the NRC implements such a rule. In December 2023, NRC staff plan to publish a proposed rule to revise the agency's radioactive source security and accountability regulations, which, if promulgated, would partially address GAO's recommendations. The final rule is not expected until June 2025.

regulations do not include specific requirements for disposing of individual sources when they are disused but still in the possession of the licensee. According to NRC officials and members of the industry we spoke with, the licensee decides when, if at all, to disposeof a source that is no longer suitable. The NRC maintains decommissioning requirements that apply when licensed activities at an entire building, facility, or specific outdoor area cease or are expected to cease, but these do not pertain to individual radioactive sources that havebecome disused.[27] This lack of regulatory direction for disused source management may contribute to some licensees delaying the disposal of certain high-risk sources, according to a report by the Disused Sources Working Group.[28]

Agency officials and licensees we spoke with recognize an elevated safety and security risk associated with continuing to possess some disused sources, but the NRC maintains that its security requirements aresufficient to address this risk. Specifically, NRC officials told us that licensees' storage of disused sources is not a safety or security concern, as licensees are subject to the same safety and security requirements as long as they possess a source, whether they are using it or not. However,the NRC has acknowledged in official documents that during long periodsof storage, sources could become lost, abandoned, or unsecured. In 2015, the NRC stated, "Licensees must store sealed sources for potentially long periods of time if there is no disposal option, and the sources are subject to loss or abandonment."[29] Furthermore, in 2014, theNRC said that "every year, thousands of sources become disused and unwanted in the United States," and "the longer sources remain disused or unwanted, the chances increase that they will become unsecured or abandoned."[30] NNSA officials also told us there is a heightened risk thatdisused sources could be stolen.[31]

---

[27] Specifically, licensees must notify the agency and either begin decommissioning or submit a decommissioning plan when (1) their license has expired; (2) they have decidedto cease activities at an entire site, building, or specific area that contains residual radioactivity in excess of NRC requirements; (3) no principal activities under the license have been conducted for 24 months; or (4) no principal activities have been conducted for 24 months in a building or specific area that contains residual radioactivity in excess ofNRC requirements. 10 C.F.R. § 30.36(d).

[28] Disused Sources Working Group, "A Study of the Management and Disposition of Sealed Sources from a National Security Perspective" (Low-Level Radioactive WasteForum, Inc.: March 2014).

[29] Concentration Averaging and Encapsulation Branch Technical Position, 80 Fed. Reg.10,165 (Feb. 25, 2015).

[30] Nuclear Regulatory Commission, "NRC Regulatory Issue Summary 2014-04: NationalSource Tracking System Long-Term Storage Indicator" (May 12, 2014).

[31] This is consistent with The 2022 Radiation Source Protection and Security Task Force Report, an interagency task force composed of the NRC, DOE, and 12 other federal departments or

The NRC has taken a step to promote the disposal of high-risk radioactive sources by initiating a rulemaking to revise its financial assurance rules to cover more radioactive sources.[32] Financial assurances require licensees to make arrangements at the time a source is obtained to cover the costs of a source's ultimate disposal.[33] However, until the revised rule is finalized, there is uncertainty as to which sources it will cover. According to the NRC, the revised financial assurance rule may apply to additional category 1 and 2 sources but may not require financial assurances for category 3 radioactive sources.[34] NRC staff told us that they have considered making category 3 sources subject to financial assurance requirements but are still analyzing whether to requirefinancial assurance for some category 3 sources.[35]

As stated above, the NRC tracks the location of category 1 and 2 sources in NSTS but does not track category 3 sources in NSTS. Therefore, it would be easier for the NRC to implement financial assurance requirements only for category 1 and 2 sources because it can more easily locate these sources. NRC staff acknowledged that not centrally tracking category 3 sources makes it more difficult to hold these licenseesaccountable for financial assurances. We previously recommended that the NRC extend its requirements for national source tracking to category 3 sources, given the risks they can pose.[36] This recommendation has not been implemented.

---

agencies, which stated that the security requirements of 10 C.F.R. Part 37 "provide reasonable assurance against the theft or loss of disused sources; however, the longer sources remain disused or unwanted, the chances increase that they will become unsecured or abandoned."

[32] The NRC's rulemaking would revise 10 C.F.R. § 30.35, "Financial Assurance and Recordkeeping for Decommissioning." If the regulation were revised to include all category 1 and 2 sources, it would substantially reduce the activity threshold for sources requiring financial assurances. For example, the cesium-137 threshold for category 2 is 27 curies, while the 10 C.F.R. § 30.35 threshold for financial assurance is 100,000 curies. For americium-241, the threshold for category 2 is 16 curies, while the 10 C.F.R. § 30.35 threshold for financial assurance is 100 curies. NRC is currently preparing the regulatory basis for the rulemaking and plans to publish the proposed rule in October 2026 and the final rule in December 2027.

[33] For example, the NRC currently requires certain licensees to obtain financial assurancesfor decommissioning costs. Allowed financial assurance mechanisms include prepayment and sureties, insurance, or other guarantee methods.

[34] National Regulatory Commission, "Staff Requirements – SECY-16-0115 – RulemakingPlan on Financial Assurance for Disposition of Category 1 and 2 Byproduct Material Radioactive Sealed Sources" (Dec. 8, 2021).

[35] For more information on the history of the NRC's efforts regarding financial assurance, see app, I.

[36] GAO-16-330.

Licensees Do Not Have an Option to Dispose of Foreign-Origin Americium- 241 Sources, and the Number of Sources Is Likely Increasing Each Year

DOE cannot accept waste containing americium-241 of foreign origin dueto DOE's determination that it does not meet the criteria for disposal at WIPP. As a result, GTCC waste from these sources has no disposal pathway.[37] As we stated above, licensees must store sealed sources for potentially long periods if there is no disposal option for them.

For this review, we could not identify the number of category 3 sources in the U.S. that have reached the point in their life cycle where they would likely be disused, because the NRC does not centrally track category 3 sources, as we recommended. However, according to NNSA estimates, there are as many as 1,000 foreign-origin category 3 americium-241 sources in the U.S.,[38] as this type of source is often used in well logging devices in the oil and gas industry. Of these foreign sources, hundreds have likely reached the end of their working life (about 15 years) and are currently disused. In addition, the only new sources currently available are from foreign suppliers because, according to NNSA documents, domestic suppliers stopped providing americium-241 in 2003.[39] Beyond new sources, according to industry representatives, some used sources can be recycled and returned to service, but it is uncertain how often this occurs or whether recycling activities could be sufficient to meet demand. See figure 3 for the life cycles of both domestic and foreign-origin americium-241 used in well logging devices.

---

[37] As stated above, GTCC waste from americium-241 sources may not be suitable for near-surface disposal and also may not meet the waste acceptance criteria for DOE's WIPP, the nation's sole geologic repository, because WIPP can only accept radioactive waste generated by atomic energy defense activities. Under certain circumstances, waste from materials used commercially may qualify as defense aste, but DOE has determined that foreign-origin sources cannot be categorized as defense waste and cannot bedisposed of at WIPP.

[38] NNSA officials based this estimate on the working life of a typical americium-241 source(10 to 15 years) and the assumption that sources produced after 2003 contained imported americium-241.

[39] According to a 2011 Los Alamos National Laboratory memorandum, DOE's Isotope Production and Distribution Program stopped selling U.S.-origin americium-241 in 2003, when DOE's inventory of americum-241 was depleted. There is currently no domestic production of americum-241, and DOE has no remaining stocks available for sale. Because of the substantial domestic need, U.S. source manufacturers have been importing Russian-made amercium-241 since at least 2003. Los Alamos National Laboratory, "Use of Foreign-Origin Radioactive Material in Sealed Sources" (Aug. 5,2011).

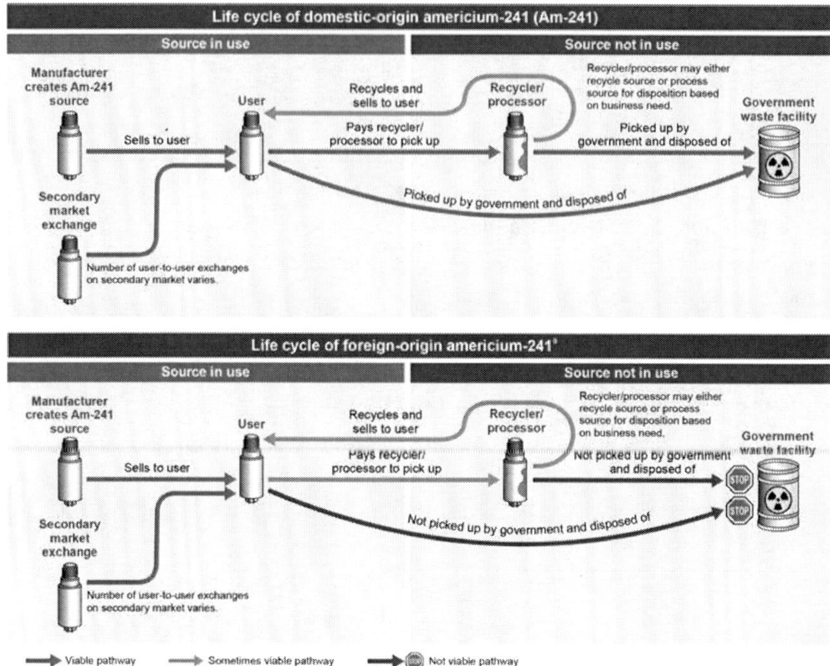

Source: GAO analysis and illustrations. | GAO-24-105998.

Note: Well logging devices are used to search for oil, gas, and water or to conduct environmental or other forms of underground monitoring. Well logging sources typically contain americium-241 and beryllium. The U.S. government produced americium-241 until the late 1980s and provided the isotope to the commercial sector until late 2003. Beginning in the early 1990s, source manufacturers started using foreign-origin americium-241 in industrial applications, and all americium-241 sources in use in the U.S. that are newer than 2003 are foreign in origin.

[a] Foreign-origin americium-241 may not be suitable for near surface disposal but also may not meet the waste acceptance criteria for the Department of Energy's Waste Isolation Pilot Plant (WIPP), the nation's sole geologic repository, because WIPP can only accept radioactive waste generated by atomic energy defense activities. Under certain circumstances, waste from materials used commercially may qualify as defense waste, but the Department of Energy has determined that foreign-origin sources cannot be categorized as defense waste and cannot be disposed of at WIPP.

Figure 3. Life Cycles of Well Logging Devices Containing Americium-241 Sources, by Their Country of Origin.

Americium-241 Source Recycling and Processing

Recycling and processing companies can offer licensees limited options for facilitating disposal or reuse of disused foreign- and domestic-origin sources. In particular, one U.S. company has the capability to recycle disused well logging devices containing americium-241.

However, the company's representative said it only recycles sources above a certain size, and a majority of the sources it recycles are of domestic origin. According to the company's representative, it may reencapsulate the disused source with a new casing. A well logging business may request this service in order to continue using the source after its casing has deteriorated. The company may also repurpose the radioactive material from the disused source in a new device of a different specification. This company also allows licensees to offload their disused sources by acting as an intermediate storage site for a nominal fee and expedite pickup by the National Nuclear Security Administration's Off-Site Source Recovery Program, according to a company representative.

The company representative said that they routinely receive requests from small well loggers that are shutting down operations and need to get rid of their disused americium-241 sources. In cases when the company will not take the source because it is of foreign origin, it sometimes facilitates the sale of the source to another well logger in the area.

Additionally, the representative told us that sources have remained with the well logging entity going out of business in a few instances when the company could not help, creating uncertainty about the ultimate disposal of these sources.

Source: GAO interview with processor. | GAO-24-105998.

Given the relatively short working life of these sources, licensees will face increasing numbers of disused category 3 foreign-origin americium-241 sources without a disposal pathway. If the NRC continues to permit licensees to possess foreign-origin americium-241 sources without a disposal pathway, the long-term security of the material could become an issue, as these sources carry an elevated risk of becoming orphaned or stolen. This is because, according to our past work, the oil and gas industry faces boom-and-bust cycles, affecting the financial stability of some well logging companies.[40] Companies that are no longer financially viable are unlikely to be able to store

[40] GAO, *Unconventional Oil and Gas Production: Opportunities and Challenges of Oil Shale Development*, GAO-12-740T (Washington, D.C.: May 10, 2012).

disused sources securely for a long period of time, increasing the risk that these sources could become orphaned, according to officials from the Disused Sources Working Group.

The NRC and DOE each have key roles in the disposal of disused sources to help mitigate the risk of these sources being used in a dirty bomb. Currently, there is no permanent disposal pathway for GTCC sources of foreign-origin americium-241 because it does not meet DOE's criteria to be disposed at WIPP and, as discussed above, DOE is required to await congressional direction before proceeding with a decision on GTCC disposal for nondefense waste. Until a permanent disposal pathway is available, finding a long-term storage solution for foreign-origin americium-241 is important for national security and will require initiative and coordination between the agencies.[41] To date, the NRC has not addressed this issue because it views disused sources as secure while they remain with the licensee, according to agency officials.

In addition, regulators—the NRC and agreement states—often do not centrally track sources containing category 3 amounts of americium-241 used in well logging devices because they are not required to, making it difficult to determine how many sources may have reached or are nearing the end of their useful lives or whether these sources are of foreign or domestic origin.[42] Further, americium-241 sources used in well logging may move around from licensee to licensee on the secondary market, and these sales may make the sources harder to keep track of. For example, two well loggers with whom we spoke said that their current stock of americium-241 sources was either inherited from a company they acquired or purchased from another well logger. Accordingly, some well loggers may be unaware if their sources are of domestic or foreign origin, which is determined by the date the source was manufactured.

Furthermore, licensees told us that the casing that surrounds americium-241 sealed sources can become worn over time, obscuring the serial number used to identify the source. Therefore, it can become more difficult to determine precisely if it is of foreign or domestic origin. (See Figure 4 of a source where the markings are beginning to wear over time.) Our analysis suggests that tracking category 3 sources would have the further benefit of

---

[41] According to the IAEA, long-term storage of a disused source means storage in a dedicated facility pending disposal. International Atomic Energy Agency, *Guidance on The Management of Disused Radioactive Sources*, Vienna, Austria (2018 ed.).

[42] The National Academy of Sciences reported in 2008 that the NRC's increased security requirements for category 1 and 2 sources prompted several oil field services companies to consider redesigning their tools to use sources just below the category 2 threshold.

improving regulators' ability to identify the locations of sources that have become orphaned and whether an americium-241 source has a disposal pathway.

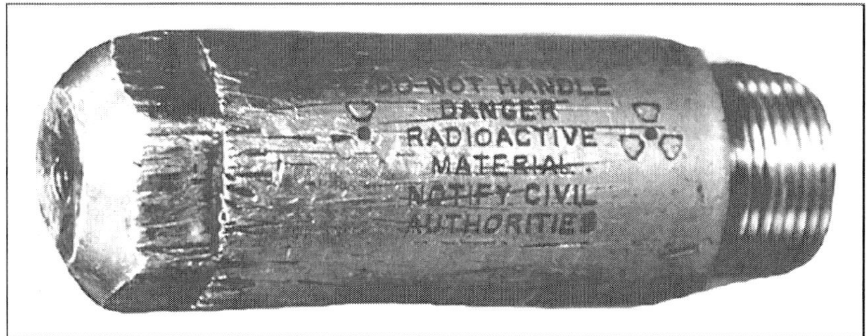

Source: Department of Energy. | GAO-24-105988.

Figure 4. Example of a Radioactive Sealed Source That Contains Americium-241.

Licensees Face a Financial Challenge Disposing of GTCC Waste from Large Cesium-137 Sources

Licensees possessing large cesium-137 sources face a financial challenge in disposing of their sources and typically rely on government subsidies to help with disposal. Large cesium-137 irradiators are likely to be considered GTCC waste and, therefore, cannot be disposed of at a commercial disposal facility. Licensees with these sources can either waitfor NNSA to pick up their material or, if possible, arrange with a radioactive materials broker—contracted by licensees—to manage the transportation of the disused source to a secure consolidation facility where DOE can assume ownership of the material prior to disposal at federal facilities. However, it may cost $200,000 to $220,000 to dispose ofwaste from a category 2 quantity of cesium-137, according to a broker.

Furthermore, some licensees we spoke with said they were unaware of disposal options and their costs when acquiring these sources. NRC doesnot require all licensees to set aside funds for disposal when it issues a license. However, some agreement states, IAEA, the Disused Sources Working Group, and the Interagency Working Group on Financial Assurances require or recommend such financial assurances.[43]

Given the challenges licensees face in disposing of their cesium-137 irradiators, NNSA often subsidizes the disposal of these sources through its Off-Site Source Recovery Program and Cesium Irradiator Replacement Project. According to NNSA officials, the agency's projectedspending for fiscal year 2022 was almost $59 million to dispose of and replace irradiators with primarily cesium-137 sources, and the agency disposed of 4,273 devices containing cesium-137 sealed sources from approximately 375 recoveries between calendar years 2003 and 2022.[44]

The Off-Site Source Recovery Program currently takes on disposal responsibilities because of the national security risks these sources pose, according to agency officials, as well as to compensate both for the lack of GTCC disposal options and for licensees' inability to pay for disposal. According to NNSA officials, the program could be streamlined if the private sector was able to take on more financial responsibility for disposition. This lack of commercial disposal facilities for GTCC waste includes larger cesium-137 sources.

Given the lack of regulatory requirements to dispose of larger cesium-137 sources, and the lack of a disposal pathway for some, the licensees may store them indefinitely or delay disposal until NNSA picks them up. This increases the risk that disused cesium-137 sources could stay in the disused state for an extended period and raises the potential that theycould become orphaned. For example:

- A licensee at a medical facility told us that their former employer could not dispose of a disused source due to the cost, and they cited security concerns at the facility where it is stored.
- An official at a medical facility told us they were not confident that they could have disposed of the facility's cesium-137 irradiators without NNSA's financial assistance. They had no idea how much money the disposal would have cost if the facility had to do it on its own. They predicted that without NNSA's financial assistance, the facility would have kept all its cesium-137 sources and accepted the

[43] The NRC has initiated a rulemaking to revise its financial assurance requirements. Therevised rule may apply to additional category 1 and 2 sources. The NRC plans to publishthe final rule in December 2027.

[44] This data was provided by NNSA and includes recoveries and disposals throughNovember 2022.

risk associated with possessing them, including the extra security measures.

- A licensee told us that without financial support from NNSA, their institution would probably still have a cesium-137 irradiator. They estimated that it may have cost the hospital $150,000 to $200,000 to dispose of it itself. The hospital had no long-term plan for disposing of the source when it purchased the irradiator, according to the licensee.

Wider Implementation of Leading Practices Could Help Address Some Disposal Challenges

Based on our reviews of key organizations' policies and guidance, we identified leading practices that could help address some disposal challenges that are not reflected in NRC requirements. These practices have been identified by the CRCPD, the Disused Sources Working Group,[45] the IAEA,[46] the Interagency Working Group on Financial Assurances,[47] the Radiation

---

[45] The Disused Sources Working Group is an independent nonprofit organization that was established to facilitate state and compact implementation of low-level waste policy and to promote the objectives of low-level radioactive waste regional compacts.

[46] The IAEA promotes management of radioactive sources through their life cycle, which includes management of disused sources. IAEA guidance establishes a holistic approach focusing on developing national policies, laws, and regulations for managing and maintaining the security of radioactive sources throughout their entire source life cycle, including when they have reached the end of their useful life. This guidance implements the IAEA's Code of Conduct on the Safety and Security of Radioactive Sources. International Atomic Energy Agency, *Guidance on The Management of Disused Radioactive Sources*. The U.S. is among the list of nations that have demonstrated commitment to implement IAEA guidance for the management of disused sources, as of June 2023. https://nucleus.iaea.org/sites/ns/code-of-conduct-radioactive-sources/Pages/default.aspx.

[47] The Interagency Working Group on Financial Assurances was established in 2008 to assess concerns related to financial assurances for radioactive material waste. The working group is comprised of several representatives from the NRC; DOE; the U.S. Environmental Protection Agency; and ICF International, LLC, as well as representatives from state regulatory agencies. The NRC considered the working group's 2010 final reportin its subsequent activities related to financial assurances for category 1, 2 and 3 sealed sources.

Source Protection and Security Task Force,[48] and WINS.[49] Some practices have also been implemented by some agreement states, helping manage disused sealed radioactive sources in those states and thus reducing risks associated with the disused phase.

Through our review of key organizations' policies and guidance, we identified six leading practices to address maintaining control over and limiting the period when sources are in disuse.[50] Table 1 outlines leading practices identified by the IAEA, WINS, selected agreement states, interagency working groups, and CRCPD.

Table 1. Leading practices for promoting management of radioactive sources through their life cycle identified by keyentities

	Tracking of category 3 sources	Source-specific financial assurances for category 1 and 2sources only[a]	Source-specific financial assurances for category 1, 2, and 3 sources	Possession fees	Possession limits	Orphan source funds
Agreement states(selected)[b]	X	n/a	X	X	X	X
Conference of Radiation Control Program Directors (CRCPD)[c]	n/a	n/a	n/a	n/a	n/a	X
Disused Sources Working Group[d]	X	n/a	X	X	X	X
Interagency Working Group onFinancial Assurances[e]	n/a	n/a	X	X	n/a	n/a

[48] The Radiation Source Protection and Security Task Force was established by the Energy Policy Act of 2005. It is led by the NRC and includes representatives from various federal agencies. According to NRC officials, the task force acts as a vehicle to coordinateand address ongoing challenges involving end-of-life management of category 1 and 2 radioactive sources.

[49] World Institute for Nuclear Security, A WINS International Best Practices Guide: Security Management of Disused Radioactive Sources (Vienna, Austria: April 2020).

[50] To identify these leading practices, we reviewed documentation and interviewed officialsfrom six agreement states, one industry group, two governmental organizations, two international organizations, and one nonprofit.

	Tracking of category 3 sources	Source-specific financial assurances for category 1 and 2sources only[a]	Source-specific financial assurances for category 1, 2, and 3 sources	Possession fees	Possession limits	Orphan source funds
International AtomicEnergy Agency (IAEA)[f]	X	n/a	X	n/a	X	X
Radiation Source Protection and Security Task Force[g]	n/a	X	n/a	n/a	n/a	n/a
World Institute forNuclear Security (WINS)[h]	X	n/a	X	n/a	X	X

Legend: X = implemented, n/a = not applicable.

Source: GAO analysis of documents from expert organizations and interviews with agreement state officials. | GAO-24-105998.

[a] The U.S. Nuclear Regulatory Commission (NRC) has initiated a rulemaking to revise its financial assurance requirements. The revised rule may apply to additional category 1 and 2 sources. The NRC plans to publish the final rule in December 2027.

[b] Agreement states assume, and the NRC discontinues, regulatory authority over specified radioactive materials. We collected information from officials from the following agreement states: Colorado, Florida, Illinois, New Jersey, Oregon, and Texas.

[c] CRCPD is a 501(c)(3) nonprofit nongovernmental professional organization dedicated to radiation protection.

[d] The Disused Sources Working Group is an independent nonprofit organization that was established to facilitate state and compact implementation of low-level waste policy and to promote the objectives of low-level radioactive waste regional compacts.

[e] The Interagency Working Group on Financial Assurances was established in 2008 to assess concerns related to financial assurances for radioactive material waste. The working group is comprised of several representatives from the NRC; the Department of Energy; the U.S Environmental Protection Agency; and ICF International, LLC, as well as representatives from state regulatory agencies. The NRC considered the working group's 2010 final report in its subsequent activities related to financial assurance for category 1, 2 and 3 sealed sources.

[f] The IAEA is the world's central intergovernmental forum for scientific and technical cooperation in the nuclear field. It promotes the safe, secure, and peaceful uses of nuclear science and technology.

[g] The Radiation Source Protection and Security Task Force was established by the Energy Policy Act of 2005. It is led by the NRC and includes representatives from various federal agencies. According to NRC officials, the task force acts as a vehicle to coordinate and address ongoing challenges, including end-of-life management of category 1 and 2 radioactive sources.

[h] WINS is an international organization whose mission is to improve professionalism and competence in nuclear security worldwide.

These six leading practices reflect four general themes. Adoption of these leading practices may more broadly incentivize timely disposal, potentially reduce overall cost to the government from orphan sources, and reduce the risk that radioactive sources could be used in a dirty bomb. These four themes are (1) source tracking, (2) financial assurances, (3) possession time limits and fees, and (4) orphan source funds.

- *Tracking category 3 sources may help the NRC and agreement states ensure licensees maintain responsibility for the security and disposal of these sources.* As we previously noted, the NRC requires tracking for category 1 and 2 sources only through the NSTS.In contrast, IAEA guidance encourages establishing a national sourceregistry to track category 3 sources as well, as we have recommended.[51]
  We identified one agreement state that has taken the initiative to trackcategory 3 sources throughout their life cycle through internal systems. Specifically, officials from Oregon told us they are able to compare state-maintained inventories to licensee records during periodic inspections. These officials said this information also helps them promote disposal of disused sources through other controls, such as applying possession time limits and imposing financial fees for disused sources held by licensees.
  According to the IAEA, having a national registry to track sources is key to ensuring comprehensive cradle-to-grave management of sources. These same benefits could accrue to category 3 sources if tracked and facilitate the use of other tools, such as establishing financial assurances.
- *Financial assurances may ensure that licensees plan for disposal of certain sources at the time of purchase.* Financial assurances require licensees to make arrangements to cover the costs of a source's ultimate disposal, which licensees told us can behigh. According to WINS, financial assurance approaches may varyaccording to the regulatory environment.[52] One type of financial assurance could be money set aside and used for disposal when a source becomes disused. Another type of financial assurance could be an upfront refundable surcharge designed to incentivize disposal (i.e., the funds are returned to the licensee when a source is disposedbut may not

[51] GAO-16-330.

[52] World Institute for Nuclear Security, A WINS International Best Practices Guide.

cover the cost of disposal). At a minimum, according to the Disused Sources Working Group, financial assurances may provide (1) some incentive for licensees to limit the time disused sources are kept before disposition and (2) a disincentive from abandoning the source. The IAEA also recommends establishing financial assurance requirements, and we identified one agreement state, Illinois, which already requires financial assurances.[53] However, as discussed above, the NRC does not specifically require disposal of disused sources, and the agency does not yet require financial assurances for all sealed radioactive sources.[54] Without such assurances, planning for disposal becomes an afterthought, resulting in delays in disposal and reliance on the federal government to pay for disposal of some radioactive sources, according to the DSWG.

Establishing requirements for financial assurances at the time of acquisition could (1) increase awareness of total life cycle costs, (2) help licensees make informed tradeoff decisions on whether to purchase and use nonradioisotopic alternative technologies,[55] (3) mitigate delays in disposal due to cost avoidance by licensees, and (4) lower licensees' reliance on the federal government to pay for disposal.

- *Possession time limits and fees for disused sources may encourage licensees to promptly dispose of these sources.* IAEArecommends minimizing the time that disused sealed sources remain in decentralized storage in order to reduce the risk of a source becoming orphaned or stolen. In contrast, the NRC does not requirelicensees to dispose of radioactive sources unless a licensee is terminating all activities under its license at specific locations.

  Some agreement states have implemented measures that may minimize the time licensees hold onto sources they no longer use. Forexample, Texas, Florida, and Oregon have implemented a rule that generally requires licensees to dispose of sources that have not been used in a 24-month period or to justify why they need to keep

[53] Illinois officials noted that their financial assurance requirements cover most category 3 sources.

[54] As discussed, the NRC's financial assurance rulemaking may require financial assurances for more category 1 and 2 sources but may not require financial assurancesfor category 3 sources.

[55] Replacing technologies that use dangerous radioactive materials with safer, nonradioisotopic alternatives may help protect people and reduce potential socioeconomic costs, if the material was released through a dirty bomb. GAO-22-104113.

them, according to state officials. Officials at the Texas Department of State Health Services told us that exemptions may be granted on a case- by-case basis, if properly justified. Additionally, officials in Oregon toldus that they impose annual fees for radioactive sources in the possession of licensees beyond the regular licensing fees. The Disused Sources Working Group reported that the fees could be an incentive for licensees to dispose of sources not in use.

- *Orphan source funds provide resources to properly dispose of lost or abandoned sources.* The IAEA recommends establishing funding mechanisms to dispose of orphaned sources and, according to NRC officials, the agency provides a grant to CRCPD to recover orphaned sources nationwide.[56] In addition, as stated above, NNSA, through its Off-Site Source Recovery Program and Cesium Irradiator Replacement Project picks up and disposes of certain large, high-risk sources. However, officials from CRCPD told us that funding is limited. Beyond this, we identified three agreement states—Texas, Illinois, and Florida–that have established their own funds that come from licensing fees. For example, officials from the Texas Departmentof State Health Services told us that having an orphaned source fund has been useful. However, agreement state officials told us that thesefunds are limited. For example, Texas officials told us that the amountof money in its fund might not be enough if they were to encounter large orphaned sources or had to dispose of multiple sources at once.

None of these leading practices has yet been adopted nationwide. The NRC has said in response to our previous recommendation that it will not track category 3 sources, but it is currently considering requiring financial assurances for more category 1 and 2 sources. Because the NRC's position is that disused sources are secure so long as they remain with the licensee, it has not assessed, in a comprehensive way, how it could better incentivize disposal of dangerous disused sources by utilizing leading practices recommended by key entities. Not planning for disposalis a potential national security risk, according to NNSA officials. In addition, a recent national security memo states that to reduce the threatof radiological terrorism, the U.S. should permanently dispose of, or recycle, disused and unwanted high-activity radioactive sources.

---

[56] As of April 30, 2023, NRC officials told us the agency has spent $83,928 on the currentCRCPD grant, with a grant ceiling of $150,000 over the life of the grant.

Conclusion

Radioactive sources provide a key benefit to many industrial and medical applications. However, if not properly used or secured, these sources pose a threat to public safety. For example, they could be stolen and used in a dirty bomb, which could result in billions of dollars in socioeconomic damage. Additionally, sources are used in industries like well logging that are subject to boom-and-bust cycles, which raises the potential for them to become orphaned if a company shuts down. The longer that sources are in a disused state, the higher the risk of loss and abandonment they have. Both the NRC and DOE agree that this risk canendure for a long period for those sources for which there is no disposal pathway and that have long half-lives, such as americium-241. Until a permanent disposal pathway can be agreed upon, the NRC and DOE may reduce these risks by taking action to find secure, long-term storagefor disused americium-241 sources.

Further, key entities, including agreement states, have identified or implemented leading practices that may reduce the time that sources remain disused, thereby reducing these risks. The NRC has begun to consider one of these practices—financial assurances—but has yet to take action to implement our prior recommendation to track category 3 sources. Moreover, the NRC has not assessed other leading practices, including possession time limits or fees for disused sources and orphan source funds. Doing so could provide better assurance of the safety andsecurity of these sources and potentially avoid significant socioeconomicconsequences.

Recommendations for Executive Action

We are making a total of three recommendations, including one to DOE and two to the NRC. Specifically:

The Secretary of Energy, in coordination with the NRC and in consultation with other relevant stakeholders, should conduct an analysis to evaluate options and take action to facilitate long-term storage, within agency authorities, to better secure foreign-origin americium-241 until a permanent disposal or viable recycling option is available. (Recommendation 1)

The Chairman of the NRC, in coordination with DOE and in consultation with other relevant stakeholders, should conduct an analysis to evaluate options and take action to facilitate long-term storage, within agency authorities, to better secure foreign-origin americium-241 until a permanent disposal or viable recycling option is available. (Recommendation 2)

The Chairman of the NRC should comprehensively assess leading practices that, if implemented, would minimize the time that disused sources are in a licensee's possession. These practices include financialassurances for all category 1, 2, and 3 sources; tracking of category 3 sources; possession time limits or fees for disused sources; and orphan source funds. (Recommendation 3)

Agency Comments and Our Evaluation

We provided a draft of this report to the Secretary of Energy, the Administrator of NNSA, and the Chairman of the NRC for review and comment. DOE, NNSA, and NRC provided written comments, which are reproduced in appendixes II and III, respectively, and summarized below.In addition, DOE, NNSA, and NRC provided technical comments, which we incorporated, as appropriate.

In its written comments, DOE and NNSA agreed with our first recommendation. Specifically, DOE and NNSA agreed to conduct an analysis of options to better secure foreign-origin americium-241 until a permanent disposal or viable recycling option is available, and have initially set April 30, 2024 as the target date for completing their analysis.

In its written comments, NRC generally agreed with our second recommendation and neither agreed nor disagreed with our third recommendation. Specifically, NRC asked for clarification on the second recommendation regarding long-term storage of foreign-origin americium-241. We have clarified our language to better reflect the intent of the recommendation. With regard to the third recommendation, NRC said thatit would need to conduct a regulatory analysis prior to implementing the leading practices identified in this report. As we have previously reported, socioeconomic costs totaling billions of dollars in damages could be avoided if a dirty bomb is prevented. In addition, the benefit of these leading practices has already been recognized by some agreement statesthat have proactively implemented them.

We are sending copies of this report to the appropriate congressional committees, the Secretary of Energy, the Administrator of NNSA, and the Chairman of the NRC. In addition, this report is available at no charge onthe GAO website at https://www.gao.gov.

If you or your staff have any questions about this report, please contact me at (202) 512-3841 or bawdena@gao.gov. Contact points for our Offices of Congressional Relations and Public Affairs may be found on the last page of this report. GAO staff who made key contributions to thisreport are listed in appendix IV.

Allison Bawden
Director, Natural Resources and Environment

Appendix I: Timeline of Recommendations and Actions Related to Financial Assurances for Radioactive Sealed Sources

The U.S. Nuclear Regulatory Commission (NRC) began exploring the need for financial assurances for additional radioactive sources in 2006, when the Radiation Source Protection and Security Task Force, which it chairs, issued a report recommending the evaluation of financial assurances for category 1 and 2 sources. This report was followed in 2010 by an interagency working group recommendation calling on the NRC and the federal government to adopt financial assurances for category 1, 2, and 3 sources. A further recommendation to require financial assurances for category 1, 2, and 3 sources was issued by the Disused Sources Working Group in 2014. The NRC is preparing the regulatory basis for a rulemaking to revise 10 C.F.R. § 30.35 to include additional category 1 and 2 sources and may explore the viability of requiring financial assurances for category 3 sources. The NRC plans to issue the proposed rule for public comment in October 2026 and the finalrule in December 2027. For a detailed timeline, see Table 2 below.

Table 2. Timeline of report recommendations and agency actions related to financial assurances of radioactive sealed sources

Date	Action
2006	In its inaugural report, the Radiation Source Protection and Security Task Force recommends that the U.S. Nuclear Regulatory Commission (NRC) evaluate the financial assurance required for possession of category 1 and 2 sourcesto assure that funding is available for the final disposition of these sources.
2010	An interagency working group—created to carry out the evaluation recommended in the 2006 task force report— recommends that the NRC and the federal government adopt a variety of financial assurance models for category 1, 2,and 3 sources.
2014	The Radiation Source Protection and Security Task Force recommends that the NRC evaluate the need for radioactivesource licensees to address the eventual disposition costs of category 1 and 2 sources through source disposition financial planning or other mechanisms.[a]
2014	The Disused Sources Working Group recommends that the NRC develop financial assurance requirements for alllicensees of category 1, 2, and 3 sources.
2016	NRC staff recommends that the agency expand its financial assurance requirements to include all category 1 and 2 sources and requests Commission approval to initiate a rulemaking.[b]
2021	The Commission approves the staff's recommendation to initiate a rulemaking to expand the agency's financial assurance requirements to include all category 1 and 2 sources. As part of this rulemaking, the Commission directs the staff to consider and seek public comment on whether financial assurance requirements should also be extended tocategory 3 sources.
2022/2023	The NRC is preparing the regulatory basis to support a rulemaking revising its financial assurance requirements for radioactive materials in 10 C.F.R. § 30.35. Staff told us they expect the rulemaking to address category 1 and 2 sources and that they are still exploring the viability of covering category 3 sources. The NRC plans to publish the finalrule in December 2027.

Source: GAO analysis of regulations and documents from expert organizations and interagency working groups. | GAO-24-105998.

[a] Elsewhere in the 2014 report, the task force states the recommendation a bit differently, saying that the NRC should "formally consider supplementing existing requirements" for licensees of category 1 and 2 sealed sources to address disposition costs.

[b] The staff noted that they believe the most prudent use of resources would be for category 1 and 2 sources, which present the highest risk. The staff also said that agreement states could continue to implement more comprehensive financial assurance requirements for sources, including category 3, based on current compatibility categories. Staff were directed by the Commission to consider and seek public comment on whether financial assurance requirements should also be extended to category 3 sources.

Appendix II: Comments from the Department of Energy and the National Nuclear Security Administration

Department of Energy
Under Secretary for Nuclear Security
Administrator, National Nuclear Security Administration
Washington, DC 20585

November 6, 2023

Ms. Allison B. Bawden
Director, Natural Resources
and Environment
U.S. Government Accountability Office
Washington, DC 20548

Dear Ms. Bawden:

Thank you for the opportunity to review the Government Accountability Office (GAO) draft report, *High-Risk Radioactive Material: Opportunities Exist to Improve the Security of Sources No Longer in Use* (GAO-24-105998). The Department of Energy's National Nuclear Security Administration (DOE/NNSA) appreciates GAO's review on the disposition of radioactive materials and shares the auditors' view on the importance of securing high-risk sources that are no longer in use with limited or no permanent disposal pathway.

DOE/NNSA agrees with the auditors' recommendation for DOE/NNSA, in coordination and consultation with the Nuclear Regulatory Commission (NRC) and other relevant stakeholders, to conduct an analysis of options to better secure foreign-origin americium-241 until a permanent disposal or viable recycling option is available. The initial target for completing this analysis is April 30, 2024. Future actions to facilitate any identified long-term storage options will be coordinated with the NRC and other relevant stakeholders.

DOE/NNSA subject matter experts have provided technical and general comments under separate cover for your consideration to enhance the clarity and accuracy of the report. If you have any questions about this response, please contact Dean Childs, Director, Audits and Internal Affairs, at (202) 836-3327.

Sincerely,

Jill Hruby

Appendix III: Comments from the Nuclear Regulatory Commission

UNITED STATES
NUCLEAR REGULATORY COMMISSION
WASHINGTON, D.C. 20555-0001

October 27, 2023

Allison B. Bawden, Director
Natural Resources and Environment
U.S. Government Accountability Office
441 G Street NW
Washington, DC 20548

SUBJECT: U.S. NUCLEAR REGULATORY COMMISSION COMMENTS ON DRAFT GOVERNMENT ACCOUNTABILITY OFFICE REPORT GAO-24-105998, "HIGH-RISK RADIOACTIVE MATERIAL: OPPORTUNITIES EXIST TO IMPROVE THE SECURITY OF SOURCES NO LONGER IN USE"

Dear Director Bawden:

Thank you for the opportunity to review and comment on the United States Government Accountability Office (GAO) draft report GAO-24-105998, "High-Risk Radioactive Material: Opportunities Exist to Improve the Security of Sources No Longer in Use," which the U.S. Nuclear Regulatory Commission (NRC) received on September 28, 2023.

We are in general agreement with the findings and first two recommendations. However, we understand the second recommendation to mean that NRC should analyze long-term storage options, which is under our authority. GAO should clarify the second recommendation such that it cannot be read to imply that NRC would be involved in the construction and operation (i.e., implementation) of long-term storage. We understand the intent behind the third recommendation; however, we would need to conduct a regulatory analysis to assess whether there is sufficient safety and security benefit to justify the burden associated with the recommended actions. Tracking of Category 3 sources is not consistent with prior Commission direction, which considered costs and benefits. The staff also has comments on certain statements made in the draft report, which can be found in the enclosure.

If you have any questions concerning the staff's comments, please contact John Jolicoeur. Mr. Jolicoeur can be reached at (301) 415-1642 or by email to John.Jolicoeur@nrc.gov.

Sincerely,

Signed by Dorman, Dan on 10/27/23

Daniel H. Dorman
Executive Director
for Operations

Enclosure:
As stated

Appendix IV: GAO Contact and Staff Acknowledgments

GAO Contact

Allison Bawden at (202) 512-3841 or bawdena@gao.gov

Staff Acknowledgments

In addition to the contact named above, Ned Woodward (Assistant Director), Jeffrey Barron (Analyst in Charge), William Bauder, Mark Braza, Antoinette Capaccio, Lilia Chaidez, Craig Comen, Lidiana Cunningham, Scott Henderson, Dan Royer, Caitlin Scoville, and Whitney Starr made key contributions to this report.

GAO's Mission

The Government Accountability Office, the audit, evaluation, and investigative arm of Congress, exists to support Congress in meeting its constitutional responsibilities and to help improve the performance and accountability of the federal government for the American people. GAO examines the use of public funds; evaluates federal programs and policies; and provides analyses, recommendations, and other assistance to help Congress make informed oversight, policy, and funding decisions. GAO's commitment to good governmentis reflected in its corc values of accountability, integrity, and reliability.

Obtaining Copies of GAO Reports and Testimony

The fastest and easiest way to obtain copies of GAO documents at no cost is through our website. Each weekday afternoon, GAO posts on its website newlyreleased reports, testimony, and correspondence. You can also subscribe to GAO's email updates to receive notification of newly posted products.

Order by Phone

The price of each GAO publication reflects GAO's actual cost of production anddistribution and depends on the number of pages in the publication and whetherthe publication is printed in color or black and white. Pricing and ordering information is posted on GAO's website, https://www.gao.gov/ordering.htm.

Place orders by calling (202) 512-6000, toll free (866) 801-7077, or TDD (202) 512-2537.

Orders may be paid for using American Express, Discover Card, MasterCard,Visa, check, or money order. Call for additional information.

Connect with GAO

Connect with GAO on Facebook, Flickr, Twitter, and YouTube. Subscribe to our RSS Feeds or Email Updates. Listen to our Podcasts.Visit GAO on the web at https://www.gao.gov.

To Report Fraud, Waste, and Abuse in Federal Programs

Contact FraudNet:

Website: https://www.gao.gov/about/what-gao-does/fraudnet Automated answering system: (800) 424-5454 or (202) 512-7700

Congressional Relations

A. Nicole Clowers, Managing Director, ClowersA@gao.gov, (202) 512-4400, U.S.Government Accountability Office, 441 G Street NW, Room 7125, Washington, DC 20548

Public Affairs

Chuck Young, Managing Director, youngc1@gao.gov, (202) 512-4800

U.S. Government Accountability Office, 441 G Street NW, Room 7149 Washington, DC 20548

Strategic Planning and External Liaison

Stephen J. Sanford, Managing Director, spel@gao.gov, (202) 512-4707
U.S. Government Accountability Office, 441 G Street NW, Room 7814, Washington, DC 20548.

Index

A

B

C

D

E

G

I

L

M

N

Y